T0202440

Lecture Notes
in Business Information Processing 395

Series Editors

Wil van der Aalst ⓘ
  RWTH Aachen University, Aachen, Germany
John Mylopoulos ⓘ
  University of Trento, Trento, Italy
Michael Rosemann ⓘ
  Queensland University of Technology, Brisbane, QLD, Australia
Michael J. Shaw
  University of Illinois, Urbana-Champaign, IL, USA
Clemens Szyperski
  Microsoft Research, Redmond, WA, USA

More information about this series at http://www.springer.com/series/7911

Mohamed Anis Bach Tobji ·
Rim Jallouli · Ahmed Samet ·
Mourad Touzani · Vasile Alecsandru Strat ·
Paul Pocatilu (Eds.)

Digital Economy

Emerging Technologies and Business Innovation

5th International Conference on Digital Economy, ICDEc 2020
Bucharest, Romania, June 11–13, 2020
Proceedings

Springer

Editors
Mohamed Anis Bach Tobji ⓘ
University of Manouba
Manouba, Tunisia

Rim Jallouli ⓘ
University of Manouba
Manouba, Tunisia

Ahmed Samet ⓘ
Institut National des Sciences Appliqué
Strasbourg, France

Mourad Touzani ⓘ
NEOMA Business School
Mont-Saint-Aignan, France

Vasile Alecsandru Strat ⓘ
Bucharest Business School
Bucharest, Romania

Paul Pocatilu
Bucharest University of Economic Studies
Bucharest, Romania

ISSN 1865-1348   ISSN 1865-1356   (electronic)
Lecture Notes in Business Information Processing
ISBN 978-3-030-64641-7   ISBN 978-3-030-64642-4   (eBook)
https://doi.org/10.1007/978-3-030-64642-4

© Springer Nature Switzerland AG 2020
This work is subject to copyright. All rights are reserved by the Publisher, whether the whole or part of the material is concerned, specifically the rights of translation, reprinting, reuse of illustrations, recitation, broadcasting, reproduction on microfilms or in any other physical way, and transmission or information storage and retrieval, electronic adaptation, computer software, or by similar or dissimilar methodology now known or hereafter developed.
The use of general descriptive names, registered names, trademarks, service marks, etc. in this publication does not imply, even in the absence of a specific statement, that such names are exempt from the relevant protective laws and regulations and therefore free for general use.
The publisher, the authors and the editors are safe to assume that the advice and information in this book are believed to be true and accurate at the date of publication. Neither the publisher nor the authors or the editors give a warranty, expressed or implied, with respect to the material contained herein or for any errors or omissions that may have been made. The publisher remains neutral with regard to jurisdictional claims in published maps and institutional affiliations.

This Springer imprint is published by the registered company Springer Nature Switzerland AG
The registered company address is: Gewerbestrasse 11, 6330 Cham, Switzerland

Preface

The constant evolution of information and communication technologies has a tremendous impact on communication and information management systems and methods. Besides, it significantly contributes to the improvement of efficiency in companies, institutions, and brings more and more usefulness to users. By placing information at the heart of products, processes, and behaviors, the fields of telecommunication and computer engineering have today revolutionized business, management, as well as daily habits. This heavy technological and sociological trend gave birth to new research opportunities, original topics, and reflection themes to both academicians and practitioners.

Initiated in 2016, the International Conference on Digital Economy (ICDEc) is on its way to become a traditional friendly and instructive yearly meeting. It offers a platform, where discussions and exchanges allow enriching theory and practice related to the evolving role of information system technologies in business, management, innovation, e-commerce, and beyond.

Consistent with its core theme, the 5th edition of the ICDEc 2020, held during June 11–13, 2020, engaged more with telecommunication and computer possibilities by embracing the virtual world in which we had our conference this year. This was possible by creating a new and challenging virtual conference experience for the participants. Breaking with traditional formats, this year's edition gained in interactivity and connectedness through visio-conference tools, and pre-recorded paper presentations. The paper entitled "Holding ICDEc 2020 Fully Online: Driving Principles and Key Decisions", published in the introductory section, presents the main strategic and technological choices of the conference to make the event successful. The core theme of this year – "Emerging Technologies & Business Innovation" – gave rise to various fascinating subthemes, ranging from digital business model innovations, data analytics, man-machine interactions, interactivity, digitalization and technological change, and the processes through which these concepts/phenomena influence organizational and societal realities such as corporate culture, enterprise architecture, business models, recommendation systems, and users' online/offline presence.

To maintain what is progressively becoming an ICDEc research tradition, the papers submitted to the competitive sessions were reviewed using a double-blind peer-review process. Each paper received at least three reviews. In spite of the COVID-19 special circumstances, 13 papers were selected thanks to the substantial contribution of several distinguished reviewers – mainly PhD. researchers and full professors – in the fields of computer science and business innovation, from about 30 universities around the world. In this regard, we would like to express our thanks to these reviewers for their dedication, the richness and depth of their comments and suggestions, and their commitment to ICDEc.

The participants in this edition of the ICDEc conference appreciated the insightful keynote speakers' speeches, namely: Claude Diderich from innovate.d llc (Switzerland) who talked about "Digital Business Model Innovation towards Competitive Advantage", Florin Gheorghe Filip from the Romanian Academy - INCE and BAR (Romania) who focused on "DSS-Concepts and Enabling I&C Technologies to Enhance Problem Solving Capabilities of Decision-makers", Daniel Traian Pele from the Bucharest University of Economic Studies (Romania) who presented "A Statistical Classification of Cryptocurrencies", and Sihem Romdhani from Veeva Systems (Canada) who proposed "Embeddings for Recommendation Systems".

We would also like to express our deepest gratitude to the country chairs, the Organization and Finance Committees, as well as the Scientific and Program Committees for their support to make this conference successful. Special thanks go to the sponsors and scientific partners of the conference, mainly Bucharest Business School from the Bucharest University of Economic Studies, Romania.

The intended audience of this book will mainly consist of researchers and practitioners in the following domains: Data science and analytics, digital transformation, digital business models, digital marketing, cryptocurrency, and recommender systems.

October 2020

Mohamed Anis Bach Tobji
Rim Jallouli
Ahmed Samet
Mourad Touzani
Vasile Alecsandru Strat
Paul Pocatilu

Organization

General Chair

Paul Pocatilu   Bucharest University of Economic Studies, Romania

Program Committee Co-chairs

Ahmed Samet   INSA Strasbourg, France
Mourad Touzani   NEOMA Business School, France

Steering Committee

Mohamed Anis Bach Tobji   University of Tunis, Tunisia
Rim Jallouli   University of Manouba, Tunisia

Organization Committee Chair

Vasile Alecsandru Strat   Bucharest University of Economic Studies, Romania

Publication Chair

Meriam Belkhir   University of Sfax, Tunisia

IT Chair

Nassim Bahri   One Way IT, Tunisia

Finance Chair

Ismehene Chahbi   University of Manouba, Tunisia

Organization Committee

Zeineb Ayachi   University of Manouba, Tunisia
Teissir Benslama   University of Manouba, Tunisia
Meriam Belkhir   University of Sfax, Tunisia
Ismehene Chahbi   University of Manouba, Tunisia
Wided Guezguez   University of Tunis, Tunisia
Ilyes Manai   University of Manouba, Tunisia
Melek Meziane   Ministry of Higher Education and Scientific Research, Tunisia

Community Management

Zeineb Ayachi	University of Manouba, Tunisia
Teissir Benslama	University of Manouba, Tunisia
Afef Herelli	University of Manouba, Tunisia

Program Committee

Molka Abbès	NEOMA Business School, France
Adnan Mustafa AlBar	King Abdulaziz University, Saudi Arabia
Paulo Almeida	Polytechnic of Leiria, Portugal
Hamida Amdouni	University of Manouba, Tunisia
Stuart Barnes	King's College London, UK
Andrea Francesco Barni	University of Applied Sciences and Arts of Southern Switzerland, Switzerland
Deny Bélisle	Université de Sherbrooke, Canada
Mariem Belkhir	University of Sfax, Tunisia
Norchene Ben Dahmane	University of Carthage, Tunisia
Mohamed Aymen Ben Hajkacem	University of Tunis, Tunisia
Chiheb-Eddine Ben N'Cir	University of Manouba, Tunisia
Karim Ben Yahia	University of Tunis, Tunisia
Hatem Bibi	University of Manouba, Tunisia
Imene Boukhris	University of Tunis, Tunisia
Manuel Castro	Universidad Nacional de Educación a Distancia, Spain
Norhène Chabchoub	University of Manouba, Tunisia
Yousra Chabchoub	ISEP, France
Anis Charfi	European Business School, France
Soumaya Cheikhrouhou	Université de Sherbrooke, Canada
Fatma Choura Abida	University Tunis El Manar, Tunisia
Claude Diderich	innovate.d llc, Switzerland
Faiza Djidjekh	University of Biskra, Algeria
Costinel Dobre	West University of Timisoara, Romania
Carlos Henrique Duarte	BNDES, Brazil
Meriam Elghali	University of Manouba, Tunisia
Frank Emmert-Streib	Tampere University, Finland
Yamna Ettarres	University of Manouba, Tunisia
Frédéric Gimello-Mesplomb	University of Avignon, France
Ebru Gokalp	Baskent University, Turkey
Thabo Gopane	University of Johannesburg, South Africa
Michael Georg Grasser	Medical University of Graz, Austria
Gabriela Grosseck	West University of Timisoara, Romania
Dorra Guermazi	University of Manouba, Tunisia
Tarek Hamrouni	University Tunis El Manar, Tunisia
Payam Hanafizadeh	Allameh Tabataba'i University, Iran
Nizar Hariri	Université Saint-Joseph, Lebanon

Reaan Immelman	Absa Group Limited, South Africa
Dyah Ismoyowati	Universitas Gadjah Mada, Indonesia
Kerem Kayabay	Middle East Technical University, Turkey
Naila Khan	Birmingham City University, UK
Nora Johanne Klungseth	Norwegian University of Science and Technology, Norway
Hasna Koubaa	University of Manouba, Tunisia
Thuraya Mellah	University of Manouba, Tunisia
Kun Chang Lee	Sungkyunkwan University, South Korea
Mhamdi Chaker	Buraimi University College, Oman
Anca Milovan	West University of Timisoara, Romania
Arindam Mukherjee	Indian Institute of Management Ranchi, India
Klimis Ntalianis	University of West Attica, Greece
Nessrine Omrani	Paris School of Business, France
Ebba Ossiannilsson	ICDE, Sweden
Catalina Ovando	UPAEP University, Spain
Malgorzata Pankowska	University of Economics in Katowice, Poland
Julia Pueshel	Neoma Business School, France
Cao Qiushi	INSA Rouen, France
Elisabetta Raguseo	Politecnico di Torino, Italy
Kai Reinhardt	University of Applied Sciences for Engineering and Economics, Germany
Tahereh Saheb	Tarbiat Modares University, Iran
Deepak Saxena	Trinity College Dublin, Ireland
Etienne Schneider	University of Strasbourg, France
Hamish Simmonds	The Australian National University, Australia
Nadine Sinno	Lebanese International University, Lebanon
Hamida Skandrani	University of Manouba, Tunisia
Fatma Smaoui	University of Tunis, Tunisia
Hanlie Smuts	University of Pretoria, South Africa
Steve Tadelis	University of California, Berkeley, USA
Ines Thabet	University of Manouba, Tunisia
Soukeina Touiti	University of Tunis, Tunisia
Imène Trabelsi Trigui	University of Sfax, Tunisia
Amira Trabelsi Zoghlami	University Tunis Carthage, Tunisia
Krisztián Varga	Corvinus University of Budapest, Hungary
Sami Yangui	Institut National des Sciences Appliquées, France
Maria Carolina Zanette	NEOMA Business School, France

Organizers

ASSOCIATION TUNISIENNE D'ÉCONOMIE NUMÉRIQUE

BUCHAREST BUSINESS SCHOOL

ACADEMY OF ECONOMIC STUDIES OF BUCHAREST

Scientific and Organizing Partners

ÉCOLE SUPÉRIEURE D'ÉCONOMIE NUMÉRIQUE

UNIVERSITÉ DE LA MANOUBA

THE INTERNATIONAL UNIVERSITY OF BEIRUT

 Springer

SPRINGER

Sponsors

BUREAU DES ETUDES TECHNIQUES D'ASSISTANCE ET DE PILOTAGE

WESS E-COMMERCE

ATB

ARAB TUNISIAN BANK

Invited Talks

Digital Business Model Innovation Towards Competitive Advantage

Claude Diderich

innovate.d, Switzerland
claude.diderich@innovate-d.com

At least since the advent of Amazon, digital business model innovation is in every manager's mouth. Many successful firms, including Airbnb, Facebook, and Apple, rely on fundamentals of digital business models to define their strategy. To understand the drivers of their success, it is important to grasp what determines success. The academic answer to this generic question is also generic, namely "strategy". Strategy, according to Ansoff, Andrews, or Barney is rooted in allocating scarce resources. Strategists, like Kaplan, Norton, or Steyer, see planning as the foundation of strategic success. Porter, another renown strategy researcher, defines success as identifying and exploiting competitive advantages in an industry setting. Others, like Mintzberg, see success rooted in managerial decision making. All these definitions of success fail to consider the specificities of a digital world, including being service oriented, its ever-changing nature, the availability of big data, and the blurring of the notion of industry.

Martin defines strategy as making informed choices about how to play and win the competitive game. This definition comes closer to a digital economy reality. A core element of how to play in a digital world is defining a firm's business model. The business model framework is the new concept for leading organizational renewal in a customer-centric digital world. The simplest definition of a business model is a description of i) how a firm *creates value for its customers* and ii) how it *captures value for itself*. When analyzing business models of successful firms, five key attributes can be identified and related to answers from five key questions:

1) *Customer segments* – Who are the customer segments targeted by the firm and what are their needs to be addressed, or, in Christensen's terminology, what are the jobs that customers want to get done?
2) *Value proposition* – How do the products and services that the firm wants to sell generate value for customers by addressing their needs and supporting them to get their jobs done?
3) *Capabilities* – How does the firm ensure that it can deliver upon the promises made to its customers? This is especially relevant and challenging if the offerings are intangible, like services or relationships, as is most often the case in a digital world.
4) *Financials* – How does the firm make profit from selling its offerings to its customers? Stated differently, why should customers be willing to pay the price requested by the firm to get their needs satisfied, their jobs done?
5) *Competitive advantage* – Why should customers buy the products and services from the firm, rather than prefer those of their competitors or rely on substitutes?

To design an innovative digital business model, it is important to understand three key differences between business models of brick and mortar firms (e.g. car manufacturers, shopping malls, craftsmen) and those of business models from firms operating in a digital economy.

First, many digital business models target multiple distinct customer segments and create value by connecting them. A typical example is Uber, who connects people seeking transportation from A to B with drivers offering such transportation on a customized basis. Another example are credit card companies, connecting consumers (buying products) with stores (selling products) and banks (handling the transfer of money). The traditional concept of industry is replaced by that of ecosystem.

Second, successfully digital business models focus on a superior understanding and fulfillment of customer needs. Rather than putting technology at the forefront (e.g. blockchain or artificial intelligence) or focusing on scaling a commodity service (e.g. messenger services or payments), as many less effective digital firms try to do, success comes from helping customers get their job done. For example, Amazon addresses the need for choice and availability of books rather than trying to sell books in stock. Hilti offers access to construction tools, like drillers, on demand for specific tasks as a service rather than selling them. WeTransfer supports transferring large files from one user to another instead of selling a medium.

Third, successful digital business models address the old-fashioned competitive advantage question in a novel way. Having a differentiating trait is especially important in the digital world, as it is often all too easy to copy a digital business model. Sometimes, competitive advantage is based on a winner-takes-all approach, like in the case of Facebook. But often, having a distinct hard to imitate differentiating characteristic is key. Access to specific resources (e.g. movies available on Netflix, songs on Spotify) may be considered such a competitive advantage. Against common knowledge, differentiation may come from focusing on specific customer segments, specific jobs to be done, or even specific roles in an ecosystem, rather than trying to satisfy any need from any customer at any cost.

In summary, successful digital business models exhibit four traits:

1) *Desirable* – Successfully digital business models offer products and services that customers want to buy because they address one or more of their needs and create value for them.
2) *Feasible* – Successful digital business models ensure that they can deliver upon the promises made to their customers.
3) *Viable* – Successful digital business models offer products and services at a price that customers are willing to pay and that does allow to generate a profit.
4) *Competitive* – Successful digital business models differentiate in a way that the targeted customers prefer their products and services over those of competitors or relying on substitutes.

About the Speaker

Dr. Claude Diderich is a strategy consultant specialized in design thinking-based business model innovation. He has more than 20 years of experience in business model innovation, strategy design and implementation, products and services development, business architecture optimization, and digital transformation. Through the years as a consultant, Claude has worked with numerous firms and advised them on improving their profitability through creativity and innovation in a digital age.

Claude Diderich holds a doctor ès sciences and a masters in computer science engineering from the Swiss Federal Institute of Technology in Lausanne, a certificate of advanced studies in strategy from the University of St. Gallen, and a specialization certificate in design thinking and innovation from the Darden Business School, University of Virginia. Claude is a member of the Strategic Management Society and serves as a member of the editorial review board of the Journal of Business Models. He is the author of "Design Thinking for Strategy: Innovating Towards Competitive Advantage" (2019).

DSS: Concepts and Enabling I&C Technologies to Augment Problem Solving Capabilities of Decision-Makers

Florin Gheorghe Filip

The Romanian Academy, Bucharest, Romania
ffilip@acad.ro

Abstract. Almost six decades ago, when presenting the scientific program of Stanford Research Institute, Engelbart (1962) stated: "…By *augmenting human intellect* we mean increasing the capability of a man to approach a complex problem situation, to gain comprehension to suit his particular needs, and to derive solutions to problems". The present talk is intended to show the role of I&C (*Information and Communication*) technologies in supporting people to make ever more effective decisions that are better adapted to the current business and technological context.

The talk starts by highlighting the evolving meaning and scope of automation and the role of the human agent in the management and control system architecture (Filip 2020). Then, several paradigms of the modern enterprise and the relevant enabling ICT are reviewed.

A DSS (*Decision Support System*) is defined, in the context of management and control systems, as an anthropocentric and evolving information system which is meant to implement the functions of a human support system that would otherwise be necessary to help the decision-maker to overcome his/her limits and constraints he/she may encounter when trying to solve complex and complicated decision problems that matter (Filip, Leiviskä 2009). A historical account of DSS evolution is submitted based on the paper of Power et al. (2019).

The case of collaborative decision-making processes and the corresponding *multi-participant (group) DSS* (Filip et al. 2017; Konaté et al. 2020) are analyzed. The preliminary results of the comparative analysis of the available platforms for the case of crowdsourceing-based decision-making (Ciurea, Filip 2019) are provided.

The original main classes of DSS, namely *model-oriented* and *data-oriented* systems, were proposed by Alter (1977). Since then, the I&C technology advanced and new tools such as *AI-based tools*, *Big Data*, *Cloud* and *Mobile Computing*, *Internet of Things* have enabled new DSS generations (Filip et al. 2017). A special attention is paid in the paper to AI (*Artificial Intelligence*)-based tools. Their usage to support decision-making was foreseen by Simon (1987). The combination of rule based expert systems with mathematical models within DSS (Filip 1991) is reviewed together with the story of the DIS-PATCHER, a family of early DSS designed to be used in production control in process industries (refineries, petro-chemical plants) and related systems composed of several processing units interconnected via buffer tanks. The concepts of *intelligent DSS* (Kaklauskas 2015) and *digital cognitive agents* (Rouse, Spohrer 2018) designed to continuously augment the human agent's knowledge and problem-solving capabilities are highlighted. The limitations and debatable questions about the usage of AI-based tools are then reviewed.

In his assumptions about the future *Age of information,* Drucke'r (1967) forecast that the information would become very cheap. Now, we can notice it is not only very cheap, but also abundant, diverse, complex, valuable and shows a fast dynamic especially in emergency situations as it is the nowadays one caused by the Coronavirus. The *Big Data/Data Science* domain emerged and got traction. The *concepts drift* in data streaming (Li et al. 2019) leads to ever more complicated decision situations and problems. In a *World Economic Forum* report (WEF 2019), it is stated that "Big-data decision making is a *value driver* for impact at scale, one of differentiators that transforms how technology is implemented, how people interact with technology, and how it affects business decisions". The DSS has evolved accordingly from model-oriented solutions to data-driven ones. The open questions concerning the *Big Data* ethical issues (Nair 2020) and the new *Dataism* paradigm (Harari 2016) are discussed and a service-oriented DSS platform (Candea, Filip 2016) is eventually presented.

Selected References

Alter, S.: A taxonomy of decision support systems. Sloan Manag. Rev. **19**(1), 9–56 (1977)

Candea, C, Filip, F.G.: Towards intelligent collaborative decision support platforms. Stud. Inf. Control **25**(2), 143–152 (2016)

Ciurea, C., Filip, F.G.: Collaborative platforms for crowdsourcing and consensus-based decisions in multi-participant environments. Informatica Economică **23**(2), 5–14 (2019)

Drucker, P.F.: The manager and the moron. In: Drucker P, Technology, Management and Society: Essays by Peter F. Drucker, pp. 166–177. Harper & Row, New York (1967)

Engelbart, D.C.: Augmenting Human Intellect: A Conceptual Framework. SRI Project 3578 (1962). https://apps.dtic.mil/dtic/tr/fulltext/u2/289565.pdf

Engelbart, D.C., Lehtman, H.: Working together. Byte, pp. 245–252, December 1988

Filip, F.G.: System analysis and expert systems techniques for operative decision making. Systems Analysis Modelling Simulation, August 1991

Filip, F.G.: DSS—a class of evolving information systems. In: Dzemyda, G., Bernatavičienė, J., Kacprzyk, J. (eds.) Data Science: New Issues, Challenges and Applications. Studies in Computational Intelligence, vol. 869. Springer, Cham (2020). https://doi.org/10.1007/978-3-030-39250-5_14

Filip, F.G., Leiviskä, K.: Large-scale complex systems. In: Nof. S. (eds.) Springer Handbook of Automation. Springer Handbooks. Springer, Heidelberg (2009). http://doi-org-443.webvpn.fjmu.edu.cn/10.1007/978-3-540-78831-7_36

Filip, F.G., Zamfirescu, C.B., Ciurea, C.: Computer Supported Collaborative Decision-making. Springer, Cham (2017). https://doi.org/10.1007/978-3-319-47221-8

Harari, Y.N.: Homo Deus: A Brief History of Tomorrow. Random House (2016)

Kaklauskas, A.: Biometric and Intelligent Decision Making Support. Springer, Cham (2015). https://doi.org/10.1007/978-3-319-13659-2

Konaté, J., Zaraté, P., Gueye, A., Camilleri, G.: An ontology for collaborative decision making. In: Morais, D., Fang, L., Horita, M. (eds.) GDN 2020. LNBIP, vol. 388, pp. 179–191. Springer, Cham (2020). https://doi.org/10.1007/978-3-030-48641-9_13

Lu, J., Liu, A., Song,Y., Zhang, G.: Data-driven decision support under concept drift in streamed big data. Complex & Intelligent Systems (2019). https://doi.org/10.1007/s40747-019-00124-4

Nair, S.J.: A review on ethical concerns in big data management. Int. J. Big Data Manag. 1(1), 8-25 (2020)

Power, D.J., Heavin, C., Keenan, P.: Decision systems redux. J. Decision Syst. (2019). https://doi.org/10.1080/12460125.2019.1631683

Rouse, W.B., Spohrer, J.C.: Automating versus augmenting intelligence. J. Enterp. Transform. (2018). https://doi.org/10.1080/19488289.2018.1424059

Simon, H.: Two heads are better than one: the collaboration between AI and OR. Interfaces 17(4), 8–15 (1987)

WEF (2019) Fourth Industrial Revolution Beacons of Technology and Innovation in Manufacturing. World Economic Forum

About the Speaker

Florin Gheorghe Filip was born on 25th July 1947. He graduated in Control Engineering at TU of Bucharest in 1970 and received his PhD degree from the same university in 1982. He was elected as corresponding member of the Romanian Academy in 1991 and become full member of the Academy in 1999. During 2000-2010, he was vice-president of the Romanian Academy (elected in 2000, reelected in 2004, and 2006). In 2010, he was elected president of the "Information Science and Technology" section of the Academy (re-elected in 2015, and 2019). He was the managing director of National Institute for R&D in Informatics-ICI Bucharest (1991-1997). His main scientific interests include optimization and control of large-scale complex systems, decision support systems, technology management and foresight, and IT applications in the cultural sector. He authored/coauthored over 350 papers published in international journals (IFAC J Automatica, IFAC J Control Engineering Practice, Annual Reviews in Control, Computers in Industry, Large-Scale Systems, Technological and Economical Development of Economy-TEDE, and so on) and contributed volumes printed by international publishing houses (Pergamon Press, North Holland, Elsevier, Kluwer, Chapman & Hall and so on). He is also the author/coauthor of thirteen monographs (published by Editura Tehnică, Hermès-Lavoisier, J. Wiley & Sons, Springer) and editor/co-editor of 27 volumes of contributions (published by Editura Academiei Române, Elsevier, IEEE Computer Society, and so on). He presented invited lectures in universities and research institutes, and plenary papers at scientific conferences in Brazil, Chile, China, France, Germany, Greece, Lithuania, Poland, Portugal, Rep. of Moldova, Romania, Spain, Sweden, Tunisia, and UK. More details can be found at: http://www.academiaromana.ro/sectii/sectia14_informatica/sti_FFilip.htm.

Embeddings for Recommendation Systems

Sihem Romdhani

Veeva Systems, Toronto, Canada
romdhani.sihem@gmail.com

AI (Artificial Intelligence) technology is now transforming every industry from manufacturing and life sciences to arts. Thanks to deep learning, we are able to build very sophisticated and highly accurate machine learning models.

One of the most successful areas to have adopted AI is Digital Advertising. Recommender systems have changed the consumer marketing world and reshaped the producer-consumer relationship. These are algorithms that provide recommendations to consumers for products that are similar to their choices. Recommender systems are widely used by websites such as YouTube, Netflix, or Amazon, and they have a significant impact on consumer sales. There are different machine learning techniques for building recommender systems. The latest ones use Embeddings.

In NLP (Natural Language Processing) domain, word embeddings is a technique that uses deep neural networks to learn low-dimensional representations of words, where similar words have similar representations (i.e., embeddings). Word2vec is a method used to efficiently create word embeddings. There are two key principals behind word2vec: the meaning of a word can be inferred from its context, and words with similar meanings tend to appear in similar contexts. Hence, words can get their embeddings by looking at their neighbors (words to the left and to the right of the word to be encoded).

Skipgram is one of word2vec model architectures. It aims to train a model to predict neighboring words using the current word. By doing so, we were able to learn efficient embeddings that preserve the semantic and syntactic relationships between words.

More recently, the concept of embeddings has shown to be effective in other applications outside of NLP domain. Researchers from the Web Search, E-commerce, and Marketplace domains have realized that we can use word2vec to learn embeddings of user actions by treating sequence of user actions as context. Examples include learning representations of items that were browsed or purchased or queries and ads that were clicked. These embeddings have subsequently been leveraged for creating recommendation engines.

Companies like Airbnb, Yahoo, Alibaba, Anghami, and Spotify have all benefitted from using word2vec approach to extract insights from users' behavior. They were able to efficiently compare, search, and categorize items using embedding representations. As a result, smarter and more powerful recommendation systems have been deployed, allowing real-time personalization in search ranking. As an example, Airbnb used skipgram model for Listing Embedding (vector representations of Airbnb homes) and saw a 21% increase in Clickthrough rate (CTR) on Similar Listing carousel.

Despite all the success, building AI-base recommendation systems is hard. One of the major challenges is building systems that are robust to real-world conditions.

There is still a huge gap between building a supervised model, achieving high performance on a static test set, and shipping a valuable product resilient to conditions change. And this is because machine learning systems are not good at generalizing when the underlying data distribution changes (i.e., the input data differs too much from the data they were trained on).

Amid the Covid-19 pandemic, online shopping behavior has radically changed, throwing a wrench into many recommendation engines. Systems trained on pre-pandemic consumer behavior are showing cracks and/or deteriorated accuracy caused by the sudden change in the way people now browse, binge, and buy, according to MIT Technology Review published on May 11, 2020.

To tackle this issue, the AI community needs to implement new approaches and processes for post-deployment monitoring like building an alert system to flag changes, use human-in-the-loop deployments to acquire new labels, and assemble a robust MLOps team.

In addition to creating tremendous value, recommender systems have enormous downside risk if we do not use data carefully. Datasets are critical to AI and machine learning, and they are becoming a key driver of the economy. Building a sophisticated recommender system requires the collection of a massive amount of users' data. This data usually includes sensitive information, covering almost every aspect of people's lives. While users have nearly no control over how data they generate are used, data collection often puts personal privacy at risk. Building the foundation for a responsible data economy requires creating new technologies and business models that provide trustworthy protection and control to data owners. Approaches include secure computation through the implementation of cryptographic techniques, secure hardware usage, and the ability to audit. Besides, advances in robust learning from limited and noisy data could help build more sophisticated and resilient recommender systems without compromising privacy.

About the Speaker

Sihem Romdhani received her Master of Applied Science (MASc) degree in Electrical and Computer Engineering from the University of Waterloo-Canada in 2015. Her academic research was focused on Deep Learning for Speech Recognition. Sihem has earned multiple awards including the Tunisian Government Sponsorship for graduate studies in Canada and The University of Waterloo Scholarship in 2013.

She is currently working with Veeva Systems in Toronto as a Data Scientist, where she is building Machine Learning models for Natural Language Processing. She has led multiple projects on text parsing, sequence tagging, information extraction from unstructured text data, and sentiment analysis. She has also worked on recommendation systems using different ML algorithms including Reinforcement Learning. Sihem is very interested in AI and how to solve new and challenging problems. Throughout her education, academic research, and work in the industry, she gathered experiences and knowledge that she enjoys sharing by actively doing public presentations. Sihem has been a featured speaker at the Open Data Science Conference since 2018.

Contents

Digital Business Models

Introductory Paper from the Steering Committee

Holding ICDEc 2020 Fully Online

Driving Principles and Key Decisions

Mohamed Anis Bach Tobji[(⊠)] [iD] and Rim Jallouli [iD]

ESEN, University of Manouba, Manouba, Tunisia
{anis.bach,rimjallouli}@esen.tn

Abstract. Due to the COVID-19 context, many events decide to cancel, post-pone, or to be held fully virtual. The International Conference on Digital Economy 'ICDEc' team decided to organize a fully online edition from June 11th to 13th, 2020. This paper outlines the main solutions that ICDEc team selected in line with the conference goals and positioning. Indeed, the business model of an international online conference would require changes related to targeting strategies, value proposition, registration fees, organizational capabilities, and technical tools. Based on organizers' recommendations in published reports of conferences held fully in a virtual mode, this paper emphasizes on the importance of clarifying the driving principles that guided the conference team choices.

Keywords: Online conference · International conference organization · Business models · Online platforms

1 Introduction

The International Conference on Digital Economy (ICDEc) is an annual conference launched in April 2016. It is the first international conference that called researchers from Computer Science and Business Innovation fields to discuss issues related to digital economy. Prior to 2020, conferences were held in Carthage, Sidi BouSaid, Brest and Beirut. The fifth edition was initially planned to be hosted at Bucharest Business School from June 11th to 13th, 2020. Due to COVID-19 outbreak, the ICDEc committees had to choose between cancelling the conference, postponing it, or keeping the initial dates with a virtual organization. The conference team decided to go for a full online edition.

The primary motivation for such a decision was to honor the conference's commit-ment towards the authors: indeed, the conference team wanted to avoid any further delay related to the lengthy publication process. Furthermore, the ICDEc is positioned as the unique annual international conference on digital economy. The online version of the conference came as a response to the global restrictive COVID-19 context.

The adoption of a new business model to hold the conference fully online would require a new set of modifications related to targeting strategies, value proposition, reg-istration fees, organizational capabilities, and technical tools. These aspects are covered throughout this paper.

© Springer Nature Switzerland AG 2020
M. A. Bach Tobji et al. (Eds.): ICDEc 2020, LNBIP 395, pp. 3–10, 2020.
https://doi.org/10.1007/978-3-030-64642-4_1

2　Targeted Participants

The online format is an opportunity for researchers from over the world to participate without having to bear travel and accommodation costs. The ICDEc team seized this opportunity and invited a bigger number of experts and confirmed researchers in the field of digital transformation, Business computing, digital communication, data analytics for business, digital economy, and digital business models. The conference team minutely realized a selection of researchers' contacts on Google Scholar based on their field of research and the number of citations as an indicator of their expertise degree. Furthermore, as a recognition of their valuable contribution in the reviewing process, all members of the program committee were invited to participate as guests to all the sessions of the conference. Finally, members of the advisory board, committees 'members and chairs of previous editions were invited to enrich the discussions and chair some of the sessions. A call for participation was sent via e-mails and the social media to the large community of ICDEc including authors and participants of the last editions.

3　Revised Registration Fees

The physical organization of ICDEc's previous editions have generated substantial expenses related to travel, accommodation, coffee breaks, meals, ceremonies, gala dinners, guided cultural tours, welcoming gifts and rewards. The cost of planning an online conference is significantly lower. Therefore, registration fees for authors and participants were revised.

4　Value Proposition

The ICDEc conference aims to create a community of researchers that produces papers of high quality that discuss issues related to digital economy, emerging technologies, and business innovation. The main purpose is directed towards establishing a long-term positioning of a well-organized conference with guaranteed scientific quality. To achieve this goal, ICDEc focuses on the following criterion:

1- The reviewing process to ensure (1) the objectivity by setting a double-blind mode (2) the quality, with a program committee (PC) whose members are meticulously selected based on their publications and objective measures (such as h-index), (3) the constructive feature of the feedbacks by assigning an important number of reviewers (between 3 and 6) per paper allowing the authors to enrich their work and to open other perspectives. For this special online edition, ICDEc wanted to benefit from the online platform, and invited the conference attendees to participate in the debates of the reviewed articles, through an asynchronous tool (such as discussion forums). This process ensures ongoing discussions, even after the articles' presentation.
2- The international dimension that guided the PC members and chairs recruitment, including a worldwide attraction of works and authors from the most important laboratories and universities involved in digital economy. The international aspect of the conference enabled connections and contributed to a consistent level of synergy between international researchers, laboratories, and universities. Participation costs were significantly reduced due to the online nature of the conference. Participants had no accommodation nor transportation fees, along with low registration charges that greatly contributed to reinforcing the international dimension.

3- The high visibility, copyright protection and professional output of the publications through the Springer Lecture Notes in Business Information Processing series. In other terms, increased quality standards and indexes are applied for the published volume, through important indexing databases.

Finally, with the online format adoption, the initial dates announced through the call for papers were maintained. Subsequently, authors of selected papers published and discussed their articles on a timely manner, yet respecting safety measures.

5   A Personalized Conference Program

The confirmed speakers and participants are based in different countries: Romania, France, Tunisia, Russia, Croatia, Switzerland, Germany, Lebanon, Canada, Portugal, and South Africa. The challenge was to plan participants' interventions to fit the same program yet taking into consideration different time zones. Several simulations took place with speakers from the Middle East, Europe, and South Africa on one hand, and a keynote speaker from Toronto on another hand: On the first and second day, the program started at 8 a.m. and lasted till 2 p.m. GMT. The third day began slightly later at 9 a.m. GMT to allow the keynote speaker from Toronto to have her intervention at 12 p.m. GMT. Challenge met; the program was published in downloadable PDF format with GMT timetables. In addition, an HTML format was published on the conference's official website: https://www.aten.tn/ICDEc2020/program.html. Thanks to the efforts of the team, particularly the IT chair, the ICDEc 2020 program was displayed in a personalized way: it allowed each participant to download the program according to his time zone, and ensured that no speaker nor author missed his intervention.

6   Organizational Capabilities and Technical Tools

The steering and organization committees have experiences with e-learning and are familiar with participation in international webinars, Visio-conferences, and virtual inter-action tools. Moreover, the steering committee has set a benchmark for other conferences when dealing with decisions related to the global restrictive context of COVID 19. Published reports of conferences held fully in a virtual mode also helped ICDEc to benefit from past experiences and organizers' recommendations (Di Natale and Bolchini 2020; Gatherer et al. 2020; Virtual Conference Resource Guide & FAQs from the IEEE Computer Society 2020). One of the most complete and relevant reports was published by the EDBT/ICDT conference organization team (Bonifati et al. 2020).

The driving principles that guided ICDEc technical choices are presented in a priority rank as follows:

1- Minimizing technical problems (DP1)
2- Maximizing interactivity in a user-friendly environment (DP2)
3- Maintaining the social contact as much as possible like in a physical conference (DP3)

Table 1 summarizes the key organization decisions according to the selected driving principles of the conference. Several options and arguments are explained in Table 1 to clarify the process of decision-making followed by the steering committee.

Table 1. Organization choices according to ICDEc driving principles.

	Option 1	Option 2	Decision	Arguments	Driving principle
Organization mode	**Live (synchronous) mode** The sessions are organized live with a "face-to-face" contact, live questions, live contacts **Advantages:** Mimic physical conferences (face-to-face contact, live Q&A etc.)	**Deferred (asynchronous) mode** Presentation videos are shared. Q&A are made in text mode, and a differed mode like in discussion forums **Advantages:** - No technical risks - A mode that fits all time zones since attendees are time free to participate	**Option 1**	The basics of a conference is to gather participants from the entire world for live scientific, social, and cultural discussions. We want to maintain this important aspect in the 2020 edition and avoid a bland edition where participants interact with very structured (professional) text messages as we do in emails	**DP2 DP3**
Presentations	Live	**Recorded**	**Option 2**	The presentation of articles and key talks is the main content to discuss in the conference. Thus, it is important to avoid any technical issues with live presentations. Then, the choice is to leave the live for Q&A. Even if technical issues occur, the chat mode can take over	**DP1**
Live organization	Enabling video cameras, audio/microphone, all the sessions with participation details		**- Camera ON for all - Audio ON only for speaker - Headsets for all**	Enabling the camera for all participants is important to mimic face-to-face meetings. Seeing faces contributes greatly to the user-friendly ambiance. Using headsets for all and enabling the microphone for the speaker only reduces noise considerably	**DP2 DP3**

(*continued*)

Table 1. (*continued*)

	Option 1	Option 2	Decision	Arguments	Driving principle
Offline notes and asynchronous interaction	**ONE exclusive full live conference Advantage**: - Really mimicking a physical conference where interactions are fully live - No additional tools for the participants. All the conference takes place using one tool	**Additional tools for Offline notes and asynchronous interactions Advantages:** - Participants sometimes need to structure their ideas and questions - Authors also need time to prepare their answers with the offline possibility - Authors have an archive of discussions. It allows them to improve their work and even to initiate new research projects	**Option 2**	The participant is obliged to use at least two tools, but the advantage of an offline discussion possibility is important to be adopted in the conference. As the conference is held online, the aim is to benefit from all possible tools that serve the scientific exchanges, and even social and cultural discussions, including forums/chats	**DP1 DP2 DP3**
Duration of presentations	Regular duration as in a physical conference (20 min)	Shorter duration (15 min)	**Option 2**	In physical conferences, participants see presentations in a real environment. However, watching a screen for hours is tiring for the eyes and nocuous for concentration. Then, the format of short presentations is selected: 15mns for authors and 20mns for guest speakers	**DP2**

(*continued*)

Table 1. (*continued*)

	Option 1	Option 2	Decision	Arguments	Driving principle
Chairing session	A session chair whose role is to launch the videos, allow participants to access the room, animate the debate and present the speakers	A session chair with an exclusive scientific and chairing role. An organization member is responsible for managing all the technical staff in the session (launching videos, helping the participants to access the room etc.)	**Option 2**	As in a physical conference, the session chair focuses on chairing the session, and so fulfilling his role as facilitator to make the debate about the presented articles fluid and relevant. The organization team launches the videos, and manages all the technical issues to hold the session	**DP1** **DP2**
Social and cultural contacts	What about contacts between participants? Networking? One-to-one and even group discussion in coffee breaks?		The list of participants' contacts is available on line, so they can exchange by emails or in a chat mode where they can interact one-to-one or many-to-many	It is very important to keep the main role of a conference; allow participants to enlarge their networks, discover other cultures, and open the way for scientific collaborations etc.	**DP2** **DP3**

Based on the results in Table 1 and a thorough benchmarking of available platforms and technical tools for conferences, ICDEc organization team used 'Google Suite' to handle the different components of the conference: these include live sessions, whiteboards, offline notes and asynchronous interactions, coffee breaks and the cultural event. Table 2 summarizes implementation tools, organizational options and arguments that support each decision.

Table 2. Implementation of the organizational choices.

	Tool	Organization	Argument
Live sessions	Google Meet	One room for all the conference since we have one plenary session (no parallel ones)	- Google Suite satisfies all the requirements, with user-friendly, secured, and scalable tools
Whiteboards	Google docs and Google Jamboard	The speaker can use Google Docs or Google Jamboard to create a shared document/whiteboard and illustrate the notions he/she wants to present	- Google Suite is an integrated environment. With a unique account, participants can use all the tools to attend and enjoy the conference
Offline notes and asynchronous interaction	Google Chat	One room for each session to structure discussions and facilitate future access	- The organization team already has the Google Suite license and is familiar with its use
Coffee breaks	Google chat and Google Meet	One room for coffee breaks. Participants can create personal rooms and to launch live meetings with other participants	

7   Conclusion

Organizing a conference fully online is a fruitful learning experience. ICDEc 2020 is an international conference that focuses on emerging technologies and business innovation. It is one of the early adopters of the fully online format that copes with the COVID19 travel restrictions.

This study explains the decisions ICDEc 2020 adopted regarding targeting strategy, value proposition, registration fees, organizational capabilities, and technical tools. Furthermore, driving principles (DP) that guided ICDEc technical choices are high-lighted in this research, namely DP1- Minimizing technical problems, DP2- Maximizing interactivity in a user-friendly environment and DP3- Maintaining the social con-tact as much as possible.

This paper attempts to explain how the fifth edition of ICDEc 2020 contributed to proposing innovative solutions and how the organizational techniques added value for the conference participants and the scientific community. An example of such innovative solutions, in addition to the downloadable PDF format with GMT timetables, is a personalized HTML format that was published on the website to allow each participant to download the program directly related to the time zone.

This paper is a research aiming to serve as a starting point for further study on the value of developing new business models and organizational approaches that are more appropriate for the online format of international conferences.

References

Bonifati, A., et al.: Holding a Conference Online and Live due to COVID-19 (2020). https://arxiv.org/abs/2004.07668

Di Natale, G., Bolchini, C.: Holding conferences online due to COVID-19: the DATE experience. In: IEEE Design & Test, vol. 37, no. 3, pp. 116–118, June 2020 (2020). https://doi.org/10.1109/mdat.2020.2995140

Gatherer, A., et al.: To be or not to be—there in person: what is the future of the technical conference? In: IEEE ComSoc (2020). https://www.comsoc.org/publications/ctn/be-or-not-be-there-person-what-future-technical-conference

Virtual Conference Resource Guide & FAQs from the IEEE Computer Society (2020). https://www.computer.org/conferences/organize-a-conference/organizer-resources/hosting-a-virtual-event/cs-virtual-event-resource-guide

Digital Transformation

Corporate Culture: Impact on Companies' Readiness for Digital Transformation

Olga Stoianova(✉) ⓘ, Tatiana Lezina ⓘ, and Victoriia Ivanova ⓘ

Saint Petersburg State University, 7/9 Universitetskaya nab., Saint Petersburg 199034, Russian Federation

{o.stoyanova,t.lezina,v.ivanova}@spbu.ru

Abstract. All the models related to the assessment of companies' readiness for transformation contain the domain that specifies the level of the corporate culture. Nevertheless, there is currently no consensus on the characteristics of an "ideal" corporate culture. Moreover, it is crucial for companies planning digital transformation to understand the minimum requirements for a corporate culture that ensure the success of planned changes. The study presents the system of universal characteristics of a company's corporate culture necessary for the implementation of successful digital transformation projects. The combined use of case study and survey methods allow identifying characteristics specific to Russian companies. The distinguished characteristics are summarized as a system of criteria of companies' readiness for digital transformation. Among them are such criteria as Employee Development, Building employee loyalty to the company, Talent attraction, The existence of a knowledge-sharing system within the company, Motivation, Cross-functional collaboration, System approach to corporate culture development. Along with the metrics for characteristics evaluation, the study proposes to use evidence.

Keywords: Digital transformation · Readiness assessment · Corporate culture

1 Introduction

Digital transformation as a large-scale change affects all subsystems of the company, and the shortcoming of any of them can lead to the failure of the entire project of company's reformation. The basis and driving force of any subsystem are people. The ability to overcome existing shortcomings largely depends on their motivation, competence, and commitment to the company, which form the basis of the corporate culture. 93% of managers see the reason for failures of digital transformation projects in corporate culture; 25% of managers - in resistance and sabotage of staff.

All the models related to the assessment of companies' readiness for transformation contain the domain that specifies the level of the corporate culture. Such domains have different names, e.g. digital culture [1], digital competencies [2], digital talents [3], workforce quality [4].

What is embedded in the concept of corporate culture in the company's digital environment?

© Springer Nature Switzerland AG 2020
M. A. Bach Tobji et al. (Eds.): ICDEc 2020, LNBIP 395, pp. 13–26, 2020.
https://doi.org/10.1007/978-3-030-64642-4_2

"Organizational culture can be defined as a pattern of basic assumptions- invented, discovered or developed by a given group as it learns to cope with its problems of external adaptation and internal integration that has worked well enough to be considered valuable and, therefore, to be taught to new members as the correct way to perceive, think and feel in relation to those problems" [5]. Business Dictionary defines corporate culture as "the values and behaviors that contribute to the unique social and psychological environment of an organization" [6].

Methods and rules of doing business characterize corporate culture, as well as policies of treatment of its employees, clients, the degree of freedom in decision-making, development of new ideas and self-expression, management and feedback structure, the level of employees' perception of the company's strategy and goals. [6].

In the "era" of digitalization, the concept of corporate culture began to transform into the concept of "digital corporate culture." For companies planning digital transformation it is crucial to understand the minimum requirements for a corporate culture that ensure the success of planned changes. Though there is no standard definition of digital culture, the characteristics of this concept are widely discussed in the literature. For example, the study [7] provides the following characteristics: Innovation, Data-driven Decision-Making, Collaboration, Open Culture, Digital First Mindset, Agility and Flexibility, Customer Centricity. The MIT SMR/Glassdoor project [8] (processing of unstructured data of 500 Glassdoor client companies with subsequent ranking) distinguished the following characteristics of the corporate culture called "Big Nine Cultural Values": Agility, Collaboration, Customer-centric, Diversity, Execution, Innovation, Integrity, Performance, Respect. Even these two sets of characteristics show that there is currently no consensus on the features of an "ideal" corporate culture.

The hypothesis of the research: there are universal characteristics of the company's corporate culture that are necessary for the implementation of successful digital transformation projects, which can be generalized as a system of readiness criteria.

The purpose of the study is to distinguish universal criteria and characteristics of corporate culture that determine the readiness of companies for digital transformation, as well as metrics for their assessment.
Tasks:

1. To identify and confirm the criteria and characteristics of corporate culture that determine the readiness of companies for digital transformation
2. To propose a system of metrics to assess the characteristics of corporate culture that determine the companies' readiness for digital transformation.

The paper presents the research findings based on literary analysis, the case study, and the survey. The objects of the case study are Russian companies that have successfully implemented digital transformation projects (recognized Russian digital leaders). The survey covers Russian companies that are at different stages of digital transformation (the set of companies doesn't include the companies from the case-study).

2   Related Works

To identify the criteria for assessing corporate culture, concerning readiness for digital transformation, two groups of models were analyzed:

1. models/methods for quantitative assessment of corporate culture, confirmed by the practical cases,
2. models for assessment of readiness for the digital transformation of consulting companies, IT companies, and the academic community.

In the first stage, the objective of the study was to compare the sets of criteria for assessing corporate culture presented in the models of these two groups.

The following models represent the first group: The Denison Organizational Culture Survey - DOCS [9], Organizational Culture Assessment Instrument - OCAI [10]; Organizational Culture Inventory - OCI [11], Organizational Culture Profile – OCP [12], Organizational Culture Survey – OCS [13], Model Organizational culture diagnosis [14]. Most models of this group distinguish criteria concerning Harmonizing strategy and tactic, e.g., Harmonizing [14], Strategic emphases and Criteria of success [10], Mission [9], Outcome-oriented [12]. Also, most models define Leadership and Motivation criteria [11]. The model [10] presents the specific features of this criterion for cultures belonging to different domains. All models include the Innovation criteria, defining it as the ability to create adaptive ways to meet changing needs (Creating Change) [9], creative autonomy, and responsiveness to change [14], ability to experiment with new ideas [12]. An important criterion that determines motivation is the level of employees' understanding of their impact on the organization [13], the perception of company values, and the desire to work [14]. The level of corporate culture is also determined by Cross-functional collaboration [9] or Team-oriented cultures [12], which are manifested in the development of competencies, joint training, conflict solving [14]. All models also contain a group of criteria that characterizes Employees and Talent Development, concerning it as the opportunity to gain knowledge and develop abilities [9], improve employees [11], select professionals based on profiles, develop staff, stimulate [14].

The second group is presented by the following models/frameworks for the assessment of a company's digital transformation readiness: The Digital Business Aptitude – (Digital Talent) [3], The Digital Maturity Model (Organization & Culture) [1], The Digital Acceleration Index (Changing ways of working) [2], Leadership Digital Transformation MaturityScape (Work Source) [4], Digital IQ (Organization employees and digital culture) [15], Business Transformation Readiness Assessment (Vision, Desire, willingness and resolve to change) [16], Digital readiness and Digital reinvention (New expertise) [17], Corporate culture and Leadership [18], Innovation, Collaborative Culture [19], People and Culture [20], Organization: Roles and Responsibilities,.Collaboration, Competence [21]. Classical models of corporate culture are being developed in models of readiness assessment submitted by consulting, which quickly responds to modern market requirements. Most models for assessing the companies' readiness for digital transformation, fix the domain associated with digital culture as very important. Consulting is confident that the success of the company's transformation is determined by a Formalized strategy supported at all levels [1], Synchronization of strategies [4],

Cascading of the strategy to all management levels [15], Development and search for talents, Efforts to form professional pools [3], the Effective development and integration of internal and external talents [4]. All models define criteria related to the innovativeness of corporate culture, such as Creativity [17], Understanding and support of digital changes [4], Readiness for transformation [16], Promotion of development ideas [1]. All models identify a group of criteria that evaluates the development of personnel, including Development programs [3], Investments in digital education [1], the Transfer of value experience [4], and the Development of competencies [2], as well as Cross-functional collaboration, which is characterized by a Flexible organizational structure [4], the Quality of integration projects [1].

Academic studies of the corporate culture in the context of digital transformation distinguish such evaluation criteria as Collaboration [19–21], Leadership [18–21], Innovativeness [18–20], Talent management [18–21], Managers' involvement [18, 21].

The analysis of the models presented above makes it possible to state that the criteria of digital transformation readiness distinguished in the readiness assessment models (group 2) mostly coincide with the criteria described in the models of corporate culture assessment (group 1). Common criteria and characteristics, rated by "quantity" of their representation in the models, are presented in Table 1.

Table 1. Criteria and characteristics of corporate culture that determines readiness for digital transformation

Criteria	% of models containing the criteria (in group 2/in group 1/total)	Characteristics
Employees development	100%/100%/100%	Competence development Corporate programs of development Individual trajectories of development Transfer of experiences Use of own professionals to develop the competences of employees Mentoring System Assessment of talents from the perspective of transformation strategy System of mobile training

(continued)

Table 1. (*continued*)

Criteria	% of models containing the criteria (in group 2/in group 1/total)	Characteristics
Innovation (creativeness and initiative of employees)	33%/50%/45,5%	Creative autonomy of employees Policies/programs for promotion and implementation of employee ideas Introducing digital breakthrough ideas
Leadership involvement	100%/62,5%/62,7%	Willingness of leaders Awareness of new opportunities
Motivation and willingness to change	33%/50%/46%	Professional motivation programs Openness of employees to new technology Understanding by all participants the benefits of transformation for the organization Motivation of management
Talent management	33%/75%/64%	Attracting professionals, including through competitions, Olympiads, etc. Interaction with universities through project seminars Professional selection based on profiles
Cross-functional collaboration	33%/38%/37%	Cross-functional, inter-departmental teams Process management not linked to the organizational structure Knowledge sharing
Open culture	0/50%/37%	Integration projects with clients in priority development areas Customer experience analysis Open to external consultations

(*continued*)

Table 1. (*continued*)

Criteria	% of models containing the criteria (in group 2/in group 1/total)	Characteristics
Harmonizing strategy and tactic	100%/50%/64%	Senior executives support digital development strategy The middle and operational management level understands and supports the managers' strategy The digital strategy is clearly formalized for internal employees and external partners
Agility	33%/25%/27%	Firm aligns structure to digital transformation objectives Organizational analytics (organizational network analysis) Productivity of interaction between internal and external participants in the organization

The criteria presented in more than 30% of the models are the following: Employees Development, Leadership involvement, Talent Management, Harmonizing strategy and tactic, Motivation and willingness to change, Creativeness and Initiative of Employees, Cross-functional collaboration, Open Culture.

Conducted in the next stage, the case-study of Russian companies- leaders of transformation intended to verify these criteria and corresponding characteristics.

3   Case-Study

Sixteen companies, twelve of which are machine-building companies and other related to different areas of activity: food industry (Cherkizovo), oil and gas industry (Gazprom Neft), retail (Sportmaster), and Russian Post. The following reasons drove this choice of these companies as the research objects. Russian machine-building companies have demonstrated that it is possible to implement successful digital transformation projects despite the technological gap with foreign competitors. In this regard, the experience of these companies is particularly valuable. Companies - leaders of digital transformation in Russia from other areas were included in the study in order to test the hypothesis of the existence of universal criteria of readiness for digital transformation related to corporate culture. The evidence of the presence of the company's corporate culture characteristics, distinguished in the previous stage, was collected from open sources. Table 2 illustrates a case-study design.

Table 2. An example of the single case description

Company	Ural Optical and Mechanical Plant
Evidence	«The Lean Production Policy was developed and approved; At least 250 rationalization proposals are submitted»
Source	http://www.uomz.ru/ru/about/lean-production
Characteristic/[Metric]	Employee initiative/[Number of rationalization proposals submitted]

The results of the case study analysis concerning different criteria are listed below.

The Criteria *"Employees Development"*
Evidence concerning the criteria *"Employees Development"* is the most frequent one. Several characteristics can be distinguished within this criterion:

– organization of corporate training in corporate universities, in traditional universities (Gazprom Neft and St. Petersburg State University), in specialized research institutes (training of Moscow Machine-Building Plant "Vpered" employees at VIAM), on joint programs (Russian Helicopters Holding and United Engine Corporation);
– standardization of training programs ("the model of corporate management competencies" at Cherkizovo Group PJSC)
– availability of career development programs ("School of Leaders" at NPO Energomash, "Personnel Reserve" project at Demnichovsky Engineering Plant, "Reserve School" program at UEC-Saturn).

The Criteria *"Creativeness and Initiative of Employees"*
Implementation of the Kaizen methodology, which provides for a system of innovation proposals (e.g., in Ural Optical and Mechanical Plant) confirms the criteria *"Creativeness and Initiative of Employees."* Many companies are implementing Codes of Corporate Culture (NPO Energomash), Corporate Ethics (e.g., KAMAZ, Russian Post), and Corporate Codes (Gazpromneft), which, among other issues, require "conscious activity and initiatives of employees."

The Criteria *"Talent Management"*
Several companies prepare an external talent pool by implementing joint educational programs with schools (e.g., project Machine-Building Factory of Podolsk JSC), with universities to train students (Gazprom Neft). Some of them create programs to support young specialists (NPO Energomash), organize case championships (Gazprom Neft, Kamaz). All these activities correspond to the *"Talent Management"* criteria.

The Criteria *"Cross-functional Collaboration"*
The criteria *"Cross-Functional"* is confirmed by the presence in companies of knowledge exchange systems: in the form of information systems or platforms (e.g., in Cherkizovo Group, Gazprom Neft, Sportmaster) or mentoring systems (e.g., Council of masters at Novocherkassk Electric Locomotive Plant).

Other characteristics of corporate culture, identified in this stage, are the following:

- participation of employees in professional skill championships (Worldskills, intra-corporate competitions),
- availability of retraining programs for the older generation (Kovrov Electro-Mechanical Plant).

The case-study can't confirm some criteria "*Harmonizing strategy and tactic*", "*Leadership involvement*" and "*Motivation and willingness to change*" as all the sources of evidence are external. A survey conducted in the next stage examines the significance of these criteria for evaluation of the company's readiness for digital transformation.

4 Questionnaire

The next stage tests above hypothesis about common characteristics of corporate culture in companies. The research includes the survey analysis that covers 160 representatives of Russian companies having been at different stages of digital transformation (the set of companies doesn't include the companies from the case-study). Among respondents are both company executives and middle management of the companies from different areas: raw materials sector (8%), construction (6%), goods production (20%), industrial production (19%), trade (8%), services (33%), and ICT (6%). The questionnaire contains questions about the company's readiness for digital transformation, personnel's readiness for transformation, the company's employees' support of the digital transformation idea, employees' willingness to change their functions, organization of training and professional development processes, employees' interest in raising the level of digital competence, interaction in the company, and the level of "digitalization" of personnel's functions. The distribution of companies by the level of readiness was as follows: with a high level - 13%, with medium - 61%, with a low - 26%. The distribution of companies by the level of staff readiness was as follows: with a high level - 31%, with a medium - 61%, with a low – 46%. 23% of representatives of the surveyed companies were not able to assess the level of staff readiness.

The Chi-square test at a significance level of <0.01 allows to conclude that there is a relationship between a company's level of readiness for digital transformation and the level of personnel readiness. Figure 1 shows that 80% of respondents who rated the level of readiness of their company for digital transformation as high, also highly evaluated the level of the company's personnel readiness. However, 10% of respondents admitted that the level of readiness of personnel is low, and another 10% of respondents found it challenging to respond to the question.

Findings from the survey are listed below.

In companies with a high degree of readiness for digital transformation the average level of personnel readiness to change their functions is 4.1 points (on a five-point scale), and the average level of personnel support for digital transformation is 4.4 points; in companies with low ratings - 3.06 and 3.3, respectively. This relation confirms the criteria "*Motivation and willingness to change*" Furthermore, in 16.67% of companies with a high level of readiness for digital transformation, employees are willing to respond to

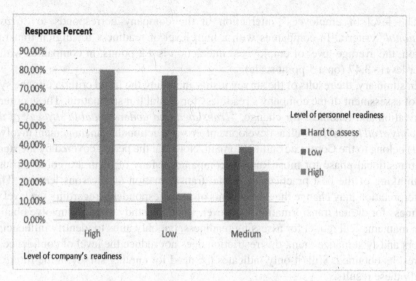

Fig. 1. Relation between personnel and the company's readiness

management's professional development proposals, in 29.4% of companies employees improve their digital competence level on their own, and in 7.2% of companies, employees apply to management for professional development proposals. In companies with a low level of readiness, these indicators are significantly lower.

The criteria *"Harmonizing strategy and tactic"* is confirmed by the fact that in 70% of companies with a high level of readiness, the goals and tasks are described in the form of strategic maps, goal trees, etc. In 62% of companies with low readiness levels, the goals and tasks are not documented. Questions on the cascading goals and the presence of KPI in the company correspond to this criterion as well. The survey showed that 46% of companies with a high readiness level use KPI, and 37% use cascading goals. In contrast, 93.5% of companies with low readiness do not use them. The results confirm the relevance of *"Harmonizing strategy and tactic"* criteria as criteria of the company's corporate culture readiness for digital transformation.

The answers concerning the organization of employee training in companies with a high level of readiness confirm the *"Employee Development"* criteria. Employees improve the qualification at least once a year in 25% of companies and improve continuously in 13% of companies. In 12.5% of companies, employees initiate the training themselves.

In our opinion, *"Leadership involvement"* criteria can be indirectly confirmed by the existence of an effective system of employee feedback. Surveys showed a high level of employee feedback implemented in various forms in companies with a high level of readiness for digital transformation. In contrast, in 38.1% of companies with low readiness for digital transformation, existing forms of feedback are formal.

The level of employees' interaction in the company corresponds to *"Cross-functional"* criteria. In companies with a high level of readiness for digital transformation, the average level of employees' interaction is 4.2 points, in companies with a lower level - 3.47 (on a 5-point scale).

In summary, the results of the survey justify including the listed criteria into the system of assessment of the company's readiness for digital transformation. These criteria (motivation and readiness for change, *"Translated and understandable strategy of the company at all levels"*, as well as development, cross-functionality, management involvement) belong to the Corporate culture domain. Note that the period covered by the study is a transitional phase for most Russian companies. Many of them are going through a rethinking of the best practices of digital transformation and lessons learned. This reinterpretation may change the assessments of some respondents regarding the level of readiness for digital transformation. However, since the study does not involve obtaining a mathematical model for assessing readiness, but only aims to identify influencing factors and systematize them, this restriction does not reduce the level of confidence in the results obtained. Still, it only indicates the need for another survey in the future to clarify these results.

5   Results

The research findings formed the basis for the proposed system of criteria, characteristics of companies' readiness for digital transformation (Table 3). Some previously distinguished criteria and characteristics are modified to match the specifics of transformation in Russian companies. Metrics for readiness assessment are recommended, as well. In some cases, evidence in binary scale can be used instead of metrics, e.g., when data required for the calculation of standard metrics are not available in the company. An evidence-based assessment model provides express readiness assessments to identify problem areas that need more detailed analysis.

Table 3. Proposals for a system of criteria/characteristics/metrics/evidence

Criteria	Characteristic/Subcharacteristic	Metrics	Evidence (1 - yes, 0 - no)
Employee development	Employee training is organized: in corporate universities/traditional universities/specialized research institutes/joint programs	For each position number of employees trained multiplied on number of hours [person*hour] or percentage of people trained [%]	Availability of appropriate form of training

(*continued*)

Table 3. (*continued*)

Criteria	Characteristic/Subcharacteristic	Metrics	Evidence (1 - yes, 0 - no)
	Training standards are being implemented	Number of professional standards are used in training planning	The company participates in the development of professional standards The company uses its own standards
	Career development programs are implemented	Number of implemented programs Proportion of program participants relative to total number of staff [%]	The company implements career development programs
Building employees engagement	Employee initiatives are encouraged	Number of innovation proposals submitted per employee per year [pcs./person] Percentage of implemented innovation proposals regarding submitted ones [%]	Lean production system is implemented in the company
	Professional skill competitions are organized	Percentage of staff participating in competitions against total number of staff [%]	The company holds professional skills competitions Employees participate in industry/Russian/international professional competitions
	Support programs for young professionals are implemented Support programs for the older generation are implemented	Number of programs with different profiles (training, housing, health, culture and leisure) Percentage of employees participating in programs [%]	The company has programs for supporting young professionals The company has programs for supporting the older generation

(*continued*)

Table 3. (*continued*)

Criteria	Characteristic/Subcharacteristic	Metrics	Evidence (1 - yes, 0 - no)
Talent attraction	Organization of Hackathons, Case Championships Collaboration with secondary education institutions Support for student studies at higher education institutions	Number of participants in such events per year [persons] Percentage of events participants who became employees of the company [%]	The company organizes events and programs to attract talents
The existence of a knowledge-sharing system (knowledge management) within the company	Functioning of information system or platform of knowledge-sharing	Average amount of information placed by the employee in the system Number of system calls per unit of time [%]	The company uses IS/platforms for knowledge sharing
	Functioning of the mentoring system	Percentage of mentors compared to the number of specialists [%] Average number of students per mentor	The company has a mentoring system
Motivation and support for transformation	Readiness for changing work functions	Percentage of employees who have changed their job functions [%]	Interviews with employees show support for the transformation
	Readiness for professional development	Percentage of employees who want to upgrade their skills [%]	
Cross-functional collaboration	Some tasks are solved by cross-functional project teams	–	Presence of cross-functional project teams

(*continued*)

Table 3. (*continued*)

Criteria	Characteristic/Subcharacteristic	Metrics	Evidence (1 - yes, 0 - no)
System approach to corporate culture development	–	–	The company adopted the Corporate Culture Code/Corporate Code

The distribution of some characteristics ("*Career development programs are implemented*", "*Professional skill competitions are organized*", "*Mentoring System*") by criteria is sometimes ambiguous. It is essential to relate such characteristics to at least one criteria of a system of the company's readiness for digital transformation assessment. For instance, "*Mentoring System*" characteristic is traditionally referred to the training field as an experience transfer system and considered as part of "Employees Development" criteria. However, in conditions when many companies rely on young employees, such a system becomes also a mechanism of maintaining intergenerational communication and appreciation of the achievements of experienced employees. It contributes to increased engagement and commitment to the company and can be referred to the "Building employee loyalty to the company" criteria.

The "*Open Culture*" criteria is not included in the proposed system as it is still difficult to verify it in Russia. For the "System approach to corporate culture development" criteria, it is difficult to distinguish any objective characteristics and metrics. An example of evidence for this criterion is the existence of formalized documents with the basic principles of corporate culture, e.g., the Code of Corporate Culture. Moreover, "*Harmonizing strategy and tactic*" can be considered as its sub-criteria.

6 Conclusion

The paper findings confirm the hypothesis about the existence of universal characteristics of a company's corporate culture necessary for the implementation of successful digital transformation projects. The combined use of case study and survey methods allow identifying characteristics specific to Russian companies. The distinguished characteristics are summarized as a system of criteria of companies' readiness for digital transformation. Among them are such criteria as Employee Development, Building employee loyalty to the company, Talent attraction, The existence of a knowledge-sharing system within the company, Motivation, Cross-functional collaboration, System approach to corporate culture development. Along with the metrics for characteristics evaluation, the study proposes to use evidence. The proposed assessment system will be tested on companies of different size and types of activity and improved in future research.

References

1. Gill, M., VanBoskirk, S.: The digital maturity model 4.0. benchmarks: digital business transformation playbook (2017). https://forrester.nitro-digital.com/pdf/Forrester-s%20Digi tal%20Maturity%20Model%204.0.pdf. Accessed 07 Feb 2020

2. Digital Acceleration Index. BCG (2016). https://www.bcg.com/ru-ru/capabilities/techno logy-digital/digital-acceleration-index.aspx. Accessed 07 Feb 2020

3. Are you ready for digital transformation? Measuring your digital business aptitude. KPMG (2016). https://assets.kpmg.com/content/dam/kpmg/pdf/2016/04/measuring-digital-business-aptitude.pdf. Accessed 07 Feb 2020

4. IDC MaturityScape: Leadership Digital Transformation. IDC (2015). https://www.idc.com/downloads/DX_UBER.pdf. Accessed 07 Feb 2020

5. Shein, E.: Organizational Culture and Leadership: A Dynamic View. Jossey-Bass Inc., San Fransisco (1985)

6. BusinessDictionary. WebFinance Inc. (2019). http://www.businessdictionary.com/definition/organizational-culture.html. Accessed 07 Feb 2020

7. The Digital Culture Challenge: Closing the Employee-Leadership Gap. Capgemini Digital Transformation Institute (2017). https://www.capgemini.com/wp-content/uploads/2017/12/dti_digitalculture_report.pdf. Accessed 07 Feb 2020

8. Sull, D., Sull, C., Chamberlain, A.: Measuring culture in leading companies. MIT Sloan Manag. Rev. Glassdoor (2019). https://sloanreview.mit.edu/projects/measuring-culture-in-lea ding-companies/#chapter-6. Accessed 07 Feb 2020

9. Denison. D.: Corporate Culture and Organizational Effectiveness. Wiley, New York (1990). https://doi.org/10.2307/258613

10. Cameron, K., Quinn, R.: Diagnosing and Changing Organizational Culture: Based on the Competing Values Framework. Jossey-Bass, San Francisco (2006)

11. Cooke, R., Lafferty, J.: Organizational Culture Inventory. Human Synergistics, Plymouth (2003)

12. O'Reilly, C., Chatman, J., Caldwell, D.: People and organizational culture: a profile comparisons approach to assessing person-organization fit. Acad. Manag. J. **34**(3), 487–516 (1991). https://doi.org/10.2307/256404

13. Glaser, S., Zamanou, S., Hacker, K.: Measuring and interpreting organizational culture. Manag. Commun. Q. **1**(2), 173–198 (1987). https://doi.org/10.1177/0893318987001002003

14. Curteanu, D., Constantin, I.: Organizational culture diagnosis - a new model. Manag. J. Faculty Bus. Adm. Univ. Bucharest **11**(1), 14–21 (2010)

15. A decade of digital Keeping pace with transformation. 2017 Global Digital IQ® Survey: 10th anniversary edition. PWC (2017). https://i40-self-assessment.pwc.de/i40. Accessed 25 Dec 2019

16. TOGAF Version 9.2. Business Transformation Readiness Assessment. The Open Group (2018). http://pubs.opengroup.org/architecture/togaf9-doc/arch/. Accessed 25 Dec 2019

17. Van Groningen, J.: Digital readiness and digital reinvention, the two sides of the digital dollar (2017). https://www.ibm.com/blogs/think/be-en/2017/03/15/two-sides-of-thedig ital-dollar/. Accessed 25 Dec 2019

18. Schumacher, A, Erol, S., Sihn, W.: A maturity model for assessing industry 4.0 readiness and maturity of manufacturing enterprises. In: Procedia CIRP, vol. 52, pp. 161–166 (2016). https://doi.org/10.1016/j.procir.2016.07.040

19. Sánchez, M., Zuntini, J.: Organizational readiness for the digital transformation: a case study research. Revista Gestão Tecnologia **18**(2), 70–99 (2018). https://doi.org/10.20397/2177-6652/2018.v18i2.1316

20. Bibby, L., Dehe, B.: Defining and assessing industry 4.0 maturity levels–case of the defence sector. Prod. Plan. Control. Manag. Oper. **29**(12), 1030–1043 (2018). https://doi.org/10.1080/09537287.2018.1503355

21. Wulf, J., Mettler, T., Brenner, W.: Using a digital services capability model to assess readiness for the digital consumer. MIS Q. Executive **16**(3), 171–195 (2017)

Digitalisation, Productivity, and Measurability of Digital Economy: Evidence from BRICS

Thabo J. Gopane[(✉)] [iD]

Department of Finance and Investment Management, University of Johannesburg, Johannesburg, South Africa
tjgopane@uj.ac.za

Abstract. Digitalisation brings along both positive disruptions and a plethora of negative economic externalities. The unresolved questions are bothersome: What is the extent of undesirable markets in digital economy, and how do measurement limitations conceal the true impact of digitalisation on global economies? The objective of this paper is to examine the influence of digitalisation on labour productivity. Methodologically, the study exploits the econometric framework of endogenous-growth model, and apply the innovative data set from the Conference Board's Total Economy Database on the BRICS case study. The findings of the study confirm the new productivity paradox of accelerated digitalisation that is not manifested in productivity growth. The outcome of this empirical study should benefit emerging market investors, technology sector industrialists, and policy makers.

Keywords: Digitalisation · Digital economy · Labour productivity · Productivity paradox

1 Introduction

The technology and economic disruptions of digitalisation are evident in various sectors of the society including education [2], agriculture [3], financial markets [4], and the broader economic linkages [5], to mention a few. There are numerous terms used to describe digital economy such as new economy, internet economy, or web economy [6]. The theme of digitalisation and electronic commerce is captured by Van Ark [1] 's definition that, "New Digital Economy … [is] the combination of mobile technology, ubiquitous access to the internet, and the shift toward storage, analysis, and development of new applications in the cloud…". That is, commercial transactions generated and operated through modern technology. In consideration of complexities involved in aggregate economic measurement, the International Monetary Fund [7] cautions against the definitions based on usage as this may easily include the whole economy, and they recommend that digital economy should be conceived as: "… the core activities of digitalisation, ICT goods and services, online platforms, and platform-enabled activities such as the sharing economy". In this context "digitalisation encompasses a wide range

© Springer Nature Switzerland AG 2020
M. A. Bach Tobji et al. (Eds.): ICDEc 2020, LNBIP 395, pp. 27–37, 2020.
https://doi.org/10.1007/978-3-030-64642-4_3

of new applications of information technology in business models and products that are transforming the economy and social interactions".

Based on their collaboration research, Oxford Economics Limited (consulting think tanks) and Huawei Technologies Company (technology supplier) reported that in 2016 the global economy's share of 15.5% was digital economy and that this share is predicted to increase to 24% in 2025 at a market value of $23 trillion [8].

The moral of the story is that digitalisation or digital economy is unquestionably beneficial and its distortions may be contagious but unequally distributed. What seems to exacerbate the risk of global digitalisation is related to measurability which leads to limited preparedness or foresight planning.

The research focus of this paper is in the area of digital adoption, and productivity benefit of digitalisation. In particular, the current study contributes to the debate of productivity paradox in the literature which may be summarised in three historical waves of digital economy.

First, in the late 1980's up to the mid-1990's there was a persistent increase in computer purchases, what Brynjolfsson [9:67] called "delivered computing power", in the U.S economy but it was not matched with comparable increase in productivity growth. This led the economics Nobel Laureate, Robert Merton Solow, to exclaim that "computers are everywhere visible, except in the productivity statistics" [10:36]. This now famous statement gave birth to the familiar term in economics namely, *productivity paradox*. Meaning that there was an unambiguous significant growth in computer technology whose economic knock-on effect was puzzlingly not reflected in productivity growth. The paradox was subsequently subjected to economic analysis by different researchers like David [11], Brynjolfsson [9], Kraemer & Dedrick [12], and Triplett [13] resulting incongruent interpretations. Second, in the mid-1990's to the early 2000's there was an internet fuelled economic growth which unlike the first, was positively correlated with increasing productivity growth and described as New Economy [14, 15]. This good news economy was short lived and disappeared with the emergence of dot.com crisis [16], and then the paradox continued. Thirdly, the current environment is neatly summarised by Van Ark [1:3] that: "Despite a rapid increase in business spending on capital and services in ICT the New Digital Economy … has not yet generated any visible improvement in productivity growth". This has come to be known as the *new* productivity paradox [6].

The puzzling negative or absence of positive association that emerges between digitalisation and labour productivity has been investigated in a number of individual countries such as the Swiss case study by Eliasson et al. [16], New Zealand by Carlawa and Oxley [17], Canada by Li [18], as well as Korea by Chung [19], among others.

In addition to national cases there are studies that evaluate digital economy at global scale (multiple countries together) such as Dimelis and Papaioannou [20], or Cooper and Xu [8]. A rare or missing research perspective is a study approach that examines digital economy in an integrated economic setting especially emerging markets, and the current study contributes towards closing this gap. A regional economic bloc, by design, provides a measure of integrated economy. This is to say that a synergy based economic environment provides an alternative environment to evaluate digital economy. Accordingly, one may conjecture that, depending on the level of economic integration and digitalisation harmony, such an environment may be more conducive to digitalisation linked productivity beneficiation. The current study looks at the case study of BRICS which is an acronym for Brazil, Russia, India, China, and South Africa. BRICS account for more than 40% of world population and nearly quarter of total world Gross Domestic Product (before COVID19 effect). BRICS has a collective work in progress regarding how to improve digital economy as well as individual initiates on digitalisation at national level. Therefore the current study should complement the BRICS digitisation learning process.

The rest of this paper will be organised as follows: Sect. 2 discusses the problem of measurement in digital economy. Section 3 explains the impact of digitalisation on productivity. Section 4 gives a report of empirical results on the relationship between digitalisation and productivity. Section 5 presents a discussion of results, while Sect. 6 concludes.

2 The Problem of Measurability and Digital Economy

There are more than one methods or variation in measuring aggregate economic activity such as gross national product (GNP), or gross domestic product (GDP). Numerically the different national accounting identities are able to reconcile back and forth effortlessly. So, while the narration in this paper will refer to GDP there will be no loss of information for alternative definitions of aggregate economic activity. GDP is the value of all final goods and services produced within a domestic economy through factors of production owned by both domestic and foreign citizens within a given period, usually a quarter or a year.

2.1 Aggravation of Existing Economic Measurement Problems

The problem of measurability related to digital economy centres on the inadequacies of existing methods of national accounting systems (see for example, Brynjolfsson and McAfee [21]. Accordingly, the weaknesses of GDP's measurement gaps are provoked and worsened in digital economy partly because GDP was originally conceived for non-digital economy. The measurement challenges that relate to digital economy are extensive but self-explanatory. Here is a practical list from IMF [7] 's description: lack of common or accepted definition of digital economy; lack of adequate sectorising or industry classification for internet platforms; adjustment errors in deflators for digital products; gaps in measuring activities of online platforms; understatement of growth and productivity; services that are self-produced, complexities of services that are volunteer-produced, or platforms-produced; limitation on measuring cross-border remittances, the need to treat data as product conflicts with current practice, the seeming arbitrariness in distinguishing digital sector from digital economy, as well as the old unsolved difficulty of whether household non-market production should be counted.

2.2 Perspectives of Digital Economy

The Organisation for Economic Cooperation and Development (OECD) [22] conceives of digital economy as an e-commerce (electronic commerce) in which measurement should focus on digitally-delivered service. This is close to Voorburg Group on Service Statistic (hereafter, Voorburg Group)'s interesting definition which focuses on the mode of trade namely, digitally-ordered, digitally-delivered, or platform-enabled transactions [23]. Figure 1 summarises the different intuitions of digital economy. The innermost, e-commerce is suggested by OECD, while the Voorburg Group recommends the second layer from the centre namely, digital transactions. The third layer from the centre appears to capture the digital economy intuition advanced by Cooper and Xu [8]. Further, although the digital economy is claimed to be birthed by internet [24] its domain has since expanded by the ongoing technical progress such as cloud computing, artificial intelligence, internet of things, mobile services, and blockchain technology, inter alia. The expansion of digital economy has made the interrogation of the existing GDP accuracy inevitable. The concern over the correctness and reasonability of the conventional system of national accounts has since broadened to scrutinise the elusive digital economy [25–27].

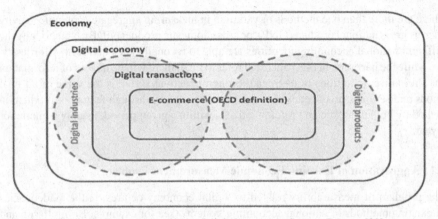

Fig. 1. Different perspectives of digital economy. Source: Barrera [23]

3   The Impact of Digitalisation on Labour Productivity

In principle the adoption of digitalisation should benefit the society (firms, workers, or consumers) through market access; ease of technology usage; innovation improvements; employment; social success; trust; and competitiveness. The sources of labour productivity improvement from digitalisation are expected to come from complementarities in technology interactions [28] for example: with firm capabilities in technical or managerial skills; innovation capacity; and efficiency promoting policies, among others. Nevertheless, the benefits of digitalisation do not seem to spread evenly across firms and industries. The potentials of digitalisation to catalyse labour productivity are desirable and feasible in numerous ways including, the improvement to product design, production processes, task automation, remote project operation, as well as the promotion of technology spillovers.

Digitalisation may also impact productivity negatively for example, the literature has observed that there is a causal correlation between excessive internet exposure and reduced sleep quality, even shirking through game playing or mobile phone obsession [29].

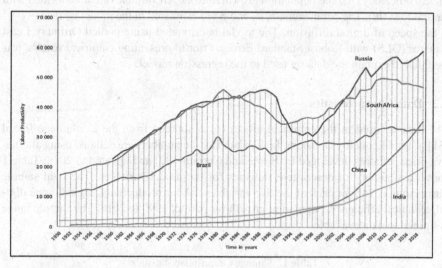

Fig. 2. Trends of labour productivity for BRICS countries. Source: Own graphics

The current paper will contribute towards understanding the anomalous relationship between digitalisation and productivity by focusing on the collective economic group of BRICS. Figure 2 shows historical trends of labour productivity from 1950 to 2018. All the BRICS countries show a steady increase in productivity historically. The upward trend was briefly interrupted by structural break in the late 1990's and early 2000's followed by a more steep upward slope which is in tandem with digitalisation growth period. Nevertheless, these are absolute numbers and further econometric analysis is essential.

3.1 Econometric Model

The goal of this empirical study is to examine the effect of digitalisation on productivity for countries in the BRICS economic bloc. Methodologically, the analysis proceeds by following Cooper and Xu [8] in applying the econometric framework of endogenous-growth model by Mankiw et al. [30]. The adapted model is presented in Eq. (1):

$$
\left[ln\left(\frac{Y_t}{L_t}\right) - ln\left(\frac{Y_0}{L_0}\right) \right] = \left(1 - e^{-\lambda t}\right)\frac{\alpha}{1 - \alpha - \beta} ln\left(\frac{I_t^D}{Y_t}\right) + \left(1 - e^{-\lambda t}\right)\frac{\beta}{1 - \alpha - \beta} ln\left(\frac{I_t^N}{Y_t}\right)
$$
$$
- \left(1 - e^{-\lambda t}\right) ln(A_0) - \left(1 - e^{-\lambda t}\right)\frac{\alpha + \beta}{1 - \alpha - \beta} ln(n) + e_t \quad (1)
$$

In this regression model the dependent variable is the change in labour productivity where Y, and L are GDP and labour quantities, respectively. The subscripts t and 0 represents time. The model has four covariates representing investment (I) relative to GDP in digital technologies $\left(\frac{I_t^D}{Y_t}\right)$, none-digital technologies $\left(\frac{I_t^N}{Y_t}\right)$, initial measure of technological efficiency (A), and population growth rate (n). All variables are transformed with natural logs (ln). The parameters, λ, α, and β are estimated in the model while $1 - e^{-\lambda t}$ is the speed of digital diffusion. The model is estimated using pooled Ordinary Least Squares (OLS) with Robust Standard Errors. Prior to presenting empirical results, it is essential to explain the data set used in the regression model.

3.2 Data Characteristics

The data set used in the empirical model (1) is sourced from the Conference Board [31]'s Total Economy Database. In this database the numbers are deflated using alternative price deflators developed by Byrne and Corrado [32] and updated to 2018. Table 1 shows a summary of descriptive statistics for the model variables. The total sample observation is 145 for each of the five variables, labour productivity, digital, none digital, technical efficiency, population growth for period, 1990 to 2018. The sample range is limited to data available.

Table 1. Summary descriptive statistics

Variable	Observation	Mean	Std Deviation	Minimum	Maximum
Labour productivity	145	−0.2388	1.8680	−3.8606	3.0000
Digital	145	4.3149	2.5677	0.0000	9.4376
None digital	145	0.8986	0.8936	−1.5000	3.7000
Technical efficiency	145	0.1765	0.7901	−2.3900	1.5200
Population growth	145	−0.0068	0.2576	−2.1717	1.6812

4 Empirical Results

Table 2 presents empirical results which are generated from the pooled OLS model (1). The statistical model was validated through the standard econometric techniques in addition to the overall fitness test through the F-statistic that satisfies the model at less than one percent level of statistical significance. Further, all the covariate coefficients are statistically significant at less than one percent level. The economic significance of the model is consistent with expectations. The positive technical efficiency may imply non-convergence (or catch up) in the sense of growth model theory [30]. The coefficient on population growth is negative as expected since an increase in population decreases labour productivity due to thin spread over a given magnitude of investment. The coefficients on both digital and none-digital technologies have a positive association with labour productivity in line with economic intuition. The implied parameter estimates, α, and β are marginal productivity for digital and none-digital technologies, respectively. Regarding digitalisation, a variable of interest in this study, the empirical results say that for a 1% increase in digital technology, labour productivity should increase by 16.91% across BRICS countries, ceteris paribus. Further intuition of the results may be illustrated through graphical interpretation.

Table 2. Empirical results

Variable	Coefficient	Robust Std Errors	P_value
Technical efficiency	1.4116	0.2420	0.0000***
Digital investment	0.2325	0.0832	0.0060***
None digital investment	0.4861	0.1637	0.0040***
Population growth	−0.9051	0.3479	0.0100***
Implied: λ α β	0.6557 0.1691 0.3537		
Constant	−1.7069	0.5121	0.0000***
F_statistic (8, 136)	43.0200		0.0000***
R_squared	71.68		
Observations	145.00		

Notes: Statistical significance at, ***1%, **5%, *10%

Figure 3 provides further explanation of the empirical relationship between digital-isation and productivity in the BRICS economic bloc. The figure shows a set of five graphs which are, estimated productivity changes generated from model (1). The graphs are plotted for the varying values of digital technology for the period of 1990 to 2018, while holding the rest of the covariates constant at their means, that is, a marginal anal-ysis. The reading of each graph shows that all the BRICS countries (with lesser extent for South Africa) reveal downward trending of productivity for the period under con-sideration. Taking the messages of both Table 2 and Fig. 3 together, we may reasonably deduce that while digital technology has positive effect on productivity it does not pro-vide enough impact for overall net increase to sustain upward trend. Further explanation and interrogation of the results is continued under the discussion section including the question of why digitalisation fail to produce an overwhelming upsurge in productivity.

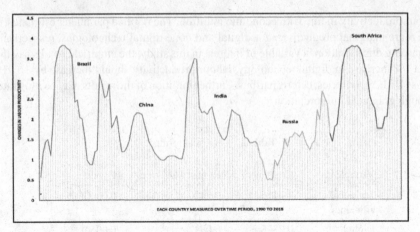

Fig. 3. Productivity responses to digitalisation in BRICS. Source: Own graphics

5 Discussion of Results

The overall research objective of this study is satisfied. The findings of the study are consistent with related studies that label the anomalous relationship between digitalisation and productivity a *new productivity paradox* [6, 18, 33]. Li [18] used firm level data for 1999 to 2005 with the generalized method of moments and confirmed the existence of productivity paradox. Both Watanabea et al. [6] and Brynjolfsson et al. [33] applied theoretical economic analysis with intuitive graphical demonstration to confirm and explain the new productivity paradox as the result of uncaptured GDP, and lagged economic effect, respectively. The latter concurs with Eliasson et al. [16] who used the micro-to-macro model to calibrate the New Economy and concluded that the first paradox of the early 1990's was a delayed economic effect or gestation rather than a paradox. Chung [19] used dynamic general equilibrium (DGE) to examine the impact of ICT on economy for the periods, 1996–2005 as well as 2006–2015 and observed labour productivity growth of 18.8% and 14.3%, respectively. Chung [19] attributed this productivity decline to economic stagnation and not paradox. This finding is an example of the myriad alternative interpretations of productivity paradox which are discussed in depth elsewhere [33]. However, in order to make the results of this study more digestible, a priority question is not why paradox occurs but: Why is it a paradox for digitalisation to fail to show a positive association with productivity growth?

In the extract below Watanabea et al. [6:230] depict a practical scenario of how modern digitalisation improves socio-economics and general wellbeing:

e-commerce as initiated by Alibaba, Amazon, and Rakuten, which sells efficiently and offers inexpensive services; search engine with online advertising such as Google and Yahoo with reduced costs for information search services; free search engines such as Wikipedia, Linux, and R with free information search and dissemination; social networks such as Twitter, Facebook, LinkedIn, and YouTube with services of finding and exchanging information efficiently; as well as cloud computing platforms such as Amazon, Apple, Cisco, IBM, Google, and Microsoft which provide services that are turning fixed costs into marginal costs.

In view of the above, common sense and economic comprehension suggest that we should expect technological progress to make work more efficient, and that this should translate into improved labour productivity (higher output per worker). On the contrary, technological improvement has not always manifested in increased labour productivity [1, 28]. In the current study, in line with similar empirical works, the evident productivity deterioration may be explained by a number of factors including, declining ICT prices, continual increase in free services in digital economy which are not captured in aggregate economic activity, as well as the known aggregate economic measurement problems. An alternative explanation of productivity paradox is that there is a lag between new technology impact on productivity and economic measurement [34, 35].

Nevertheless, a positive consolation is that the literature on digitalisation has noted ways to improve the digitalisation beneficiation towards labour productivity. For instance, policy makers and industrialists may promote digitalisation oriented labour productivity by enhancing complementarities in digitalisation: with other technologies [36]; with human skills [37], as well as with regulations, among others.

6 Conclusion

The study has reviewed the problem of measurability in digital economy and found that part of the inherited weaknesses emanates from the fact that GDP measures were originally designed for non-digital economy. Further, the paper explained the conceptual relationship between digitalisation and labour productivity followed by an empirical evaluation of the BRICS economic bloc as a case study. The findings of the study are satisfactory and confirms the presence of new technology productivity paradox observed in related studies. The results show that when digitalisation increases by 1% labour productivity should increase by 17%, other things constant. The policy implication of the research is that technology policy makers should consider improving digitalisation's beneficiation towards labour productivity by enhancing its complementarities with other assets such as applicable human skills.

References

1. Van Ark, B.: The productivity paradox of the new digital economy. Int. Prod. Monit. **31**(Fall), 3–18 (2016)
2. Gopane, T.J.: Blockchain Technology and smart universities. In: Kalpa Publications in Computing Proceedings of 4th International Conference on the Internet, Cyber Security and Information Systems, vol. 12, pp. 72–84 (2019). https://doi.org/10.29007/rkv5
3. Gopane, T.J.: What is the impact of digital financial service on agribusiness market risk? In: Cunningham, P., Cunningham, M. (eds.) IST-Africa Week Conference, Gaborone, Botswana, 9–11 May 2018, pp. 1–7. IIMC, Dublin (2018a)
4. Gopane, T.J.: The likelihood of financial inclusion in e-banking: a biprobit sample-selection modeling approach. In: Bach Tobji, M.A., Jallouli, R., Koubaa, Y., Nijholt, A. (eds.) ICDEc 2018. LNBIP, vol. 325, pp. 67–78. Springer, Cham (2018). https://doi.org/10.1007/978-3-319-97749-2_5

5. Rinaldo, E., Guerrieri, P., Meliciani, V.: The economic impact of digital technologies in Europe. Econ. Innov. New Technol. **23**(8), 802–824. https://doi.org/10.1080/10438599.2014.918438
6. Watanabea, C., Naveeda, K., Touc, Y., Neittaanmäkia, P.: Measuring GDP in the digital economy: increasing dependence on uncaptured GDP. Technol. Forecast. Soc. Chang. **137**, 226–240 (2018). https://doi.org/10.1016/j.techfore.2018.07.053
7. IMF (International Monetary Fund): Measuring the digital economy. IMF Staff report. IMF, Washington, DC (2018). https://doi.org/10.1111/roiw.12308
8. Cooper, A., Xu, W.: Digital spillover measuring the true impact of the digital economy. International Digital Economy Consultancy report: Huawei, and Economics Limited (2018)
9. Brynjolfsson, E.: The productivity paradox of information technology. Commun. ACM **36**(12), 66–77 (1993). https://doi.org/10.1145/163298
10. Solow, M.R.: Manufacturing matters: the myth of the post-industrial economy. New York Times, Book Rev. **36** (1987)
11. David, P.: The dynamo and the computer: an historical perspective on the modern productivity paradox. Am. Econ. Rev. **80**(2), 355–361 (1990). https://doi.org/10.1108/17410401011006086
12. Kraemer, K.L., Dedrick, J.: Payoffs from investment in information technology: lessons from the Asia-Pacific region. World Dev. **22**(12), 1921–1931 (1994). https://doi.org/10.1016/0305-750X(94)90183-X
13. Triplett, J.: The Solow productivity paradox: what do computers do to productivity? Can. J. Econ. **32**(2), 309–334 (1999). https://doi.org/10.2307/136425
14. Kelly, K.: New Rules for the New Economy: 10 Radical Strategies for a Connected World. Viking, New York (1998)
15. Rifkin, J.: The Third Industrial Revolution: How Lateral Power is Transforming. Energy, the Economy, and the World. Macmillan, New York (2011). https://doi.org/10.1177/1045159512467326
16. Eliasson, G., Johansson, D., Taymaz, E.: Simulating the new economy. Struct. Change Econ. Dyn. **15**, 289–314 (2004). https://doi.org/10.1016/j.strueco.2004.01.002
17. Carlawa, K.I., Oxley, L.: Resolving the productivity paradox. Math. Comput. Simul. **78**, 313–318 (2008). https://doi.org/10.1016/j.matcom.2008.01.029
18. Li, X.: An analysis of labour productivity growth in the Canadian tourism/hospitality industry. Int. J. Tour. Hosp. Res. **25**(3), 374–386 (2014). https://doi.org/10.1080/13032917.2014.882850
19. Chung, H.: ICT investment-specific technological change and productivity growth in Korea: comparison of 1996–2005 and 2006–2015. Telecommun. Policy **42**, 78–90 (2018). https://doi.org/10.1016/j.telpol.2017.08.005
20. Dimelis, P.S., Papaioannou, S.K.: Technical efficiency and the role of ICT: a comparison of developed and developing countries. Emerg. Mark. Finance Trade **47**(2), 41–54 (2015). https://doi.org/10.2307/23047100
21. Brynjolfsson, E., McAfee, A.: The Second Machine Age: Work, Progress, and Prosperity in a Time of Brilliant Technologies. W.W. Norton & Company, New York (2014)
22. OECD: Going Digital: Shaping Policies, Improving Lives. OECD Publishing, Paris (2019). https://doi.org/10.1787/9789264312012-en
23. Barrera, E., Bravo, R., Cecconi, C., Garneau, M.B., Murphy, J.: Measurement challenges of the digital economy. Policy report, 24 September 2018. Voorburg Group, Rome (2018). https://doi.org/10.1787/9789264113541-en
24. Tapscott, D.: The Digital Economy: Promise and Peril in the Age of Networked Intelligence. McGraw-Hill, New York (1994). https://doi.org/10.5465/ame.1996.19198671

25. Groshen, E.L., Moyer, B.C., Aizcorbe, A.M., Bradley, R., Friedman, D.M.: How government statistics adjust for potential biases from quality change and new goods in an age of digital technologies: a view from the trenches. J. Econ. Perspect. **31**(2), 187–210 (2017). https://doi.org/10.1257/jep.31.2.187

26. Feldstein, M.: Understanding the real growth of GDP, personal income, and productivity. J. Econ. Perspect. **31**(2), 145–164 (2017). https://doi.org/10.1257/jep.31.2.145

27. Syverson, C.: Challenges to mismeasurement explanations for the US productivity slowdown. J. Econ. Perspect. **31**(2), 165–186 (2017). https://doi.org/10.3386/w21974

28. OECD: Digitalisation and productivity: a story of complementarities. Economic Outlook report. OECD Publishing, Paris (2019b). https://doi.org/10.1787/b2e897b0-en

29. Gibson, M., Shrader, J.: Time use and labor productivity: the returns to sleep. Rev. Econ. Stat. **100**, 783–798 (2018). https://doi.org/10.1162/rest_a_00746

30. Mankiw, G., Romer, D., Weil, D.: A contribution to the empirics of economic growth. Q. J. Econ. **107**(2), 407–437 (1992). https://doi.org/10.2307/2118477

31. Conference Board: Growth Accounting and Total Factor Productivity, 1990–2018 (Adjusted version). Total Economy Database, April 2019

32. Byrne, D., Corrado, C.: ICT services and their prices: what do they tell us about productivity and technology? Finance and Economics Discussion Series, No. 2017-015. Board of Governors, Washington (2017). https://doi.org/10.17016/FEDS.2017.015r1

33. Brynjolfsson, E., Rock, D., Syverson, C.: Artificial intelligence and the modern productivity paradox: a clash of expectations and statistics. In: Agrawal, A., Gans, J., Goldfarb, A. (eds.) The Economics of Artificial Intelligence: An Agenda, pp. 23–57. University of Chicago Press, Chicago (2019). https://doi.org/10.7208/chicago/9780226613475.001.0001

34. Byrne, D., Marshall, M., Reinsdorf, B., Fernald, J.G.: Does the United States have a productivity slowdown or a measurement problem? In: Federal Reserve Bank of San Francisco Working Paper, No. 2016-03 (2016). https://doi.org/10.17016/FEDS.2016.017

35. Brynjolfsson, E., Rock, D., Syverson, C.: The productivity J-curve: how intangibles complement general purpose technologies. In: NBER Working Paper, No. 25148 (2018). https://doi.org/10.3386/w25148

36. Bartelsman, E., Van Leeuwen, G., Polder, M.: CDM using a cross-country micro moments database. Econ. Innov. New Technol. **26**, 168–182 (2017). https://doi.org/10.1080/10438599.2016.1202517

37. Bloom, N., Sadun, R., Van Reenen, J.: Americans do it better: US multinationals and the productivity miracle. Am. Econ. Rev. **102**, 167–201 (2012). https://doi.org/10.1257/aer.102.1.167

Technological Change and Future Skill-Shortages in Engineering and Architecture Education: Lessons from Lebanon

Nizar Hariri[1]([⊠]) (iD), Raymond Bou Nader[2], and Sarah Haykal[3]

[1] Faculty of Economics, OURSE, Université Saint Joseph, Beirut, Lebanon
[2] Institut de Gestion des Entreprises, Université Saint Joseph, Beirut, Lebanon
[3] Faculty of Economics, CEDREC, Université Saint Joseph, Beirut, Lebanon

Abstract. The construction sector is undergoing rapid global changes, calling for a reform in the architecture and engineering education. Lebanon is an interesting case to study skill-shortage, future occupations and career opportunities within these professions due to its booming construction sector and its high level of integration with neighboring Arab Gulf countries. An employer survey identified 3 factors - thus called "influence factors" - that are radically transforming the professions of architects and engineers: technological innovation, global outsourcing of construction services, and the organizational shift towards a more fluid work environment. This paper analyzes how these influence factors affect the employability and the future skill-gap and assesses the impact of identified changes on skill-requirements and skill-shortages. Substantive recommendations for the engineering and architecture education are suggested, including reforming curricula and competencies frameworks in order to reduce the skill-gap, with a view on their implementation.

Keywords: Engineering education · Architecture education · Skill-Shortages · Lebanon · E-Skills · Outsourcing · Creativity

1 Introduction

In the construction sector, firms have to adapt to an increasing global competition (Horns and Jenkins 2001), with an unprecedented level of outsourcing, and constantly changing legislations and rules (such as new safety norms, construction regulations, environmental concerns, etc.). These global changes are not merely exogenous constraints to which construction firms can passively adapt. Rather, there is a radical shift in the working environment, calling for thorough reforms in architecture and engineering education (Rocha Brito 2012; Lucas 2016).

With the scientific advancement in materials science, nanotechnology, or biotechnology, it is trivial to say that architecture and engineering education need to be tailored to match future jobs (Borrego and Henderson 2014). Our article is not intended to repeat such truism. It rather aims to identify the most crucial future skills for architects and engineers competing within multicultural fluid organizations. Our survey suggests

© Springer Nature Switzerland AG 2020
M. A. Bach Tobji et al. (Eds.): ICDEc 2020, LNBIP 395, pp. 38–50, 2020.
https://doi.org/10.1007/978-3-030-64642-4_4

substantive recommendations for Higher Education Institutions, with a view on their implementation.

Lebanon is an interesting case to study future occupations, career opportunities and skill-shortages within these professions due to the booming construction sector in Lebanon and its high level of integration with neighboring Arab Gulf countries. Yet, studies on the skill-mismatch are scarce. The construction sector is a major asset for the Lebanese economy, with more than 1000 firms concentrated in Beirut and its neighboring areas (Table 1). Although Lebanon is one of the smallest countries in the Middle East, the morphology of its construction sector shows a large number of big firms exceeding 200 employees, highly involved in the construction boom in neighboring regions. Moreover, Lebanese Higher Education Institutions (HEI) gave a pioneering role in the skill supply on the regional level.

Table 1. Number of firms by category in Lebanon – CCIA-BML, 2014

Firm category	Excellent	First	2nd	3rd	4th
Number of firms	107	173	178	394	308

This article follows a prospective methodology called PROMENIA, a qualitative research method that we adjusted to the Lebanese employment and labor market in order to predict future skill-needs and skill-shortages in the engineering and architecture professions[1]. The starting point of our methodology is a prospective analysis of employer's needs through interviews and focus groups with key-informants and stake-holders from the leading firms in the country. Information was gathered on current and expected employment, skill-shortages, and future skill-needs. Informants where asked to discuss different scenarios in which various factors are radically transforming their jobs.

Therefore, three "factors of influence" affecting architecture and engineering professions, calling for future educational reforms, were identified:

a) technological innovation;
b) global outsourcing of construction services; and,
c) the organizational shift toward a more fluid work environment, with more flexibility and less boundaries.

The labor market data was collected in 2014 through the registers of the Chamber of Commerce, Industry and Agriculture of Beirut and Mount-Lebanon (CCIA-BML).

---

[1] This methodology was first established and implemented by researchers from the University of Strasbourg and slightly adjusted to the Lebanese labor market, as part of the European project, PACOME, funded by the European Commission. The project was dedicated to study skill-shortage, future occupations and future career opportunities in Lebanon, between 2012 and 2016.

A pool of 1160 firms was identified, divided into five categories (excellent, 1, 2, 3, 4). In the focus group, experts were selected from the first 2 types of firms ("Excellent", meaning more than 100 employees, and "category 1" meaning 50–100 employees) for whom public information was available, in order to establish a questionnaire. This qualitative method was followed by a complementary quantitative analysis, through a semi-structured e-questionnaire submitted to 150 selected companies. Half of these companies were small firms, and the other medium-sized business and leading companies. The response rate was around 30%, with 43 questionnaires actually filled-out by Human Resources managers.

The analysis runs in three major steps. First, the morphology of the construction sector in Lebanon was scanned, in order to identify key firms and key positions and to understand how the influence factors are affecting the market structure. These morphological repercussions are referred to as the structural effect (Sect. 2). Second, we analyzed how these influence factors modify jobs, tasks and activities (which is referred to as the occupational effect), in order to build different scenarios for the evolution of future skill-needs (Sect. 3). Finally, while assessing the impact of these changes on skill-requirements and skill-shortages, major recommendations are suggested to reform curricula and competencies framework in order to reduce the skill-gap (Sect. 4).

2   The Construction Sector in Lebanon and Its Globalization

This section describes the morphology of the construction sector in Lebanon by identifying key firms and key positions within these firms, in order to analyze the structural effect, and to understand how influence factors affect the market structure.

2.1   The Morphology of the Construction Sector in Lebanon

First of all, the real estate and the construction industry occupy a major position in the Lebanese economy, since they both account for the fifth of the Lebanese GDP, on average, since 2004 (Table 2). Indeed, the Construction sector alone weighs between 4% and 6% of the GDP, while the Real estate sector varies between 12% and 14% between 2004 and 2015[2].

Table 2. Evolution of the construction sector in Lebanon (CAS 2015)

Year	2004	2005	2006	2007	2008	2009	2010	2011	2012	2013
% GDP	3,8%	3,8%	3,9%	4,4%	4,6%	5,0%	3,8%	4,5%	5,2%	6,2%

The real estate sector weighs between 14% and 16% of the GDP, between 2004 and 2013, according to National Accounts (Table 3). The accelerated growth after 2007

---

[2] In National accounts, the real growth of the sector is inferred from statistics on raw material, (local cement, and imported products used in building and public works). The variation of construction prices is estimated as a function of the weighted average of changes in the wages of manual workers and the prices of raw materials.

reflects the need for reconstruction after the 2006 war, and the rise in the price of real estate and construction materials.

Table 3. Evolution of the real estate sector in Lebanon (CAS 2015)

Year	2004	2005	2006	2007	2008	2009	2010	2011	2012	2013
Added Value	4,829	5,023	5,127	5,252	5,715	6,271	7,434	8,317	9,276	9,980
% GDP	15%	16%	15%	14%	13%	12%	13%	14%	14%	14%

The relative stagnation of the sector since 2010 and the consequent recession shows that this previous trend has reached its maturity, yet the sector remains the most important contributor to the Lebanese economic growth (Fig. 1). Despite the relative recession in the sector, it remains perfectly correlated with economic growth.

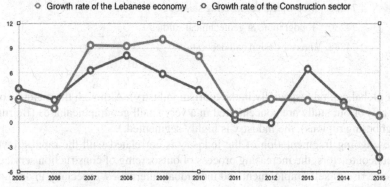

Fig. 1. Growth rates of the construction and real estate sector and the growth rate of the Lebanese GDP between 2005 and 2015 – CAS, 2019

2.2 The Structural Effect: Changes in the Market Structure

The Construction sector in Lebanon covers a wide range of activities in many different industries, such as infrastructure, building construction and development projects, demolition, wrecking, drilling or design (Table 4).

Architecture and engineering professions include a variety of skills that cover a wide range of activities including non-residential buildings, residential construction projects, civil engineering projects such as roads and bridges, mechanical engineering, interior design and decoration. Many achievements can be noticed throughout the country, not only in the reconstruction of the downtown of Beirut, but also in other sectors like lighting, waterfront marinas, and renovation of historical buildings and the built Heritage.

The Lebanese sector does not rely much on public funds, with a continuous increase in private investment in real estate, especially from 2007 till 2012, and an important

Table 4. Construction sector - registered firms (by activity) - CCIAB-ML, 2014

CODE	Activities
45110	Demolition & wrecking of buildings, earth moving
45120	Test drilling and boring for construction
45210	Construction of buildings & civil engineering works
45230	Construction of highways, roads, airfields & sport facilities
45240	Construction of water projects
45260	Construction of water systems
45270	Electrical contracting
45311	Installation of electrical wirings and fitting
45320	Installation of thermal and sound insulation in buildings
45451	Construction of swimming pools
74210	Civil engineering, architecture & Construction studies
74220	Mechanical engineering studies
74230	Topographic & geotechnical studies
74843	Interior design and decoration

flow of global capital (especially in the tourism industry). Although the majority of the 1160 firms of our study are concentrated in a very small geographical area (Beirut and its neighboring regions), the industry is highly segmented.

The ongoing fragmentation of the industry is correlated with the rapid growth of small subcontractors, the increasing process of outsourcing of construction services, as well as the rise of self-employment and individual enterprises, especially in design and animation services. Accordingly, architecture and engineering professions have evolved into a multidisciplinary structure organized around a number of corporate networks and technologically advanced globalized firms.

The Lebanese labor market is undergoing gradual structural changes in the supply and demand of skills. The supply of skills is unstable, due to the rapid increase in the number of private universities, and to dramatic political and social changes in the Arab world with higher rates of migration of the labor force. The instability in the skill supply is aggravated by global factors driven by the rise of a "new economy" based on information, knowledge, innovation and creativity (Kenner and Isaak 2004). These factors are generating skill-shortages in the construction sector, a skill mismatch between the increasing needs for more creative labor on the employment market and the outdated curricula of Higher Education Institutions.

Thus, architecture, engineering, and design are converging into one multidisciplinary field and network (Jacobides 2006). The value chain is divided into primary and support activities. Primary activities contribute to the physical creation of the product, sales or distribution, as well as after sale services. Support activities assist the primary activities but are increasingly subject to outsourcing in the architecture and engineering professions. Consequently, the Lebanese market is structured around a central hub of huge companies surrounded by networks of alliances and partnerships with small suppliers and subcontractors. The structural effect may be described as follow:

- The construction sector is organized around a huge number of small sub-contractors whose proportion continues to grow, in a market dominated by few big companies, with only 107 firms out of 1160 counting more than 100 employees.
- The segmentation of the sector and the extension of very small enterprises (700 firms with less than 10 employees) is an indicator of a high level of subcontracting.
- Various types of firms use different technologies, and highly specialized skills may be required in each one. Indeed, a lot of generic skills may be common to all the companies in the sector, but the core competencies that are strategic to each one may be specific, knowledge-based and non-transferable to other occupational types.

2.3 Structural Shifts in the Recruitment Policies

This structural effect is expected to induce gradual shifts in the recruitment policies:

- Leading firms competing on a larger scale need to hire the most creative and autonomous architects and engineers and to provide them with continuous training for sales, management and leadership (Lepak and Snell 2001).
- These corporations need to work closely with suppliers and loyal clients, through various types of alliances and partnerships, and thus, they need to identify the core competencies and the strategic skills to keep-in-house and to build-on.
- Finally, the structural effect shows that roles and functions of architects and engineers are not as easily defined as they were in the last decades, since they are increasingly following the fluid and not-well-defined tasks of managers, team leaders and negotiators.

Consequently, Higher Education Institution should adapt to the requirements of the new organizational structure.

- *Risks of the brain drain:* Indeed, globalization of engineering and architecture services is facilitated by lower cost of high-skilled labor coming from developing countries as well as by progress in education and information technologies in these countries (Arciszewski 2006). Thus, outsourcing of engineering and architecture services continues to increase, driven by the lower costs. Lebanon is already benefiting from this trend since a lot of firms based in the Gulf are currently outsourcing some services towards the Lebanese market (especially, consultant services, conceptualization and design). The backside of this trend for Lebanese firms is the increased instability of skill supply and demand on the local employment market, with a concrete risk of brain drain associated with a decrease in engineers and architects jobs. The process of outsourcing is also unstable, and the demand for Lebanese firms in the Gulf may not last. Innovation and creativity combined with the highest levels of expertise are the added value that prevents non-routine work from being lost, at least as long as Lebanese professionals maintain their competitive advantage.

– *Local vs. global trends in skill supply and Education:* The increasing demand for engineers and architects are also affected by globalization and by the use of new and more sophisticated industrial technologies. These factors may have various impacts on the Lebanese labor market, resulting in a widening gap between skill supply and demand. On one hand, the skill supply for qualified engineers and architects is mainly a domestic issue, related to higher education policies and reforms; on the other, the future needs of the market are driven by global trends that are difficult to anticipate by Lebanese universities and local policies. Thus, there might be a significant gap between what universities are expected to do and what they are currently doing in order to survive in a rapidly changing environment. Keeping in mind that an increasing number of freshly graduates are immigrating to Arab countries, especially in the Gulf, the HEI are facing contradictory exigencies: on the local scale, they are competing to attract the largest number of students through traditional standard programs focused on generic and transferable skills; on the regional scale, they have to compete with international programs in order to provide highly-skilled labor capable of embracing a wide range of new technologies and material. In conclusion, the migration of skilled professionals from the country is an increasing source of concern. The Gulf trend is a local form of globalization that seems particularly important in civil engineering and architecture because of the increasing demand for the Lebanese graduates and professionals in many design and construction companies in the Arab world. Similarly, Lebanese firms should learn to compete globally (or regionally) in order to survive.

3   The Occupational Effect: New Skills and Tasks

The previous section identified three factors that modify the structure of the construction sector: globalization, new technologies and the increase need for flexibility and creativity in the new corporate model. This section aims to study the occupational effect, in order to understand how the three identified "factors of influence" change the skill-demand in the architecture and engineering professions. In our survey, technology was considered by 65% of the respondents as the major factor influencing the profession, and it was associated with higher productivity and better quality of services. Moreover, 47% of the respondents considered that intercultural and environmental concerns radically transform their business by increasing the future need for creative architects and engineers in leadership position, especially for female professionals. Similarly, 37% of the respondents considered that these cultural changes in the working environment will lead to the emergence of unprecedented fields of expertise. This section analyzes how architecture and engineering professions are affected by these major changes, and which skills are crucial for future occupations.

3.1   Technology and the Emergence of New Occupations and Tasks

First, since some engineering and design services may be carried out remotely, technology is increasing the opportunities of outsourcing with lower costs. Secondly, new software programs are reducing the transaction costs of projects, and the time required to accomplish engineering analysis. Internet and e-mails are increasing the possibilities of interactions around the globe, and with other counterparts inside the firm environment. As a result, firms have access to a more diversified architecture and engineering staff from different educational background. In parallel, clients have become more connected with the suppliers, and project partnerships, both among and with clients, are more common. Finally, technology is transforming the regulatory framework, introducing new standards and norms, with more concern for the environmental impact of the project. With new technologies, new specializations have emerged, creating new opportunities and more jobs, especially in program management and specialty subcontractors. The growing reliance on technology has increased the demand for communication and management skills, as well as for leadership skills among architects and engineers as shown in the list of tasks that are common to both professions:

- To protect the public's health, safety and security;
- To understand and to respect regulations, local practices and international standards;
- To control and to execute parts of a project, including design criteria, analysis methods, and material selection, in collaboration with co-workers and teams;
- To communicate and to discuss with colleagues and teams, with various stakeholders, especially legal advisors and public sector officials;
- To participate in collaborative work with different teams, and to communicate architecture and/or engineering requirements and expectations, in order to achieve the project goals or solve technical problems;
- To create and evaluate alternatives in order to provide cost-effective, sustainable solutions, adjusting each project to the clients' needs
- To minimize the environmental impacts of projects, while satisfying the cultural, legal, and environmental standards.
- To supervise or to prepare plans, specifications, and reports;
- To address questions related to decisions and requirements, acting as a creative authority during the planning and the design phases, while acting as a technical authority during the construction phases;
- To initiate and maintain collaborations with key engineers, architects from different departments or other partner/alliance companies, as well as officials in order to facilitate negotiation and consultation.

The tasks required from engineers and architects in the new corporate model include some soft and generic attributes that go beyond technical achievement and excellence. The demand for these generic attributes will keep growing since the globalization of services and the technological changes shape work environment, restructuring it into a more fluid and collaborative network.

3.2   New Tasks for Architects and Engineers in Leading Positions

As more collaborative tasks are required from engineers or architects, firms are less likely to invest in training for routine activities, or tasks requiring generic skills that might be transferable from one firm to another. Leadership skills, communication skills, and managerial skills cannot be considered anymore as merely administrative supplement to the profile of a successful engineer or architect. Rather, they seem to be the core values of the profession as well as strategic inputs for the value chain. Therefore, the profile of engineers and architects in leadership positions requires more specific skills, which need to be adjusted to each particular context, and leading companies are in need of a greater focus on leadership skills, especially:

- To apply analytical skills from math and science to concrete specific and unique situations;
- To reach optimal solutions with the lowest cost and the most feasible and efficient way;
- To apply organizational development skills and managerial skills that are located in a "more abstract" set of interpersonal skills in order to enhance cooperation and coordination within each project team and between different teams;
- To identify and develop specific qualities or attitudes required for long-term leadership and to encourage autonomy and creativity among his/her co-workers.

4   Skill-Mismatch and Educational Reforms

The structural and occupational effects generate a skill-mismatch in the engineering and architecture professions. The following section addresses the employability crisis and the major challenges for education and training systems in Lebanon.

Firstly, it is important to note the stagnation of the construction sector in Lebanon has led to an important decrease in the number of students in the Engineering and architecture education.

Table 5. Enrollment in the civil engineering and architecture education in Lebanon. Source: Statistical bulletin 2019 - Center of Educational Research and Development, Ministry of Education

Year	Civil engineer students	Architecture students	Total
2016–2017	5082	3769	8851
2017–2018	5390	3682	9072
2018–2019	4347	3006	7353
Ratio (2019/2017)	−14.46%	−20.24%	−16.92%

4.1   Future Skills Needed in a Fluid and Flexible Work Environment

Most of the pioneering firms in the construction sector in Lebanon rely on well-known, confirmed and well-established professionals, and seem to pay higher salaries in order to attract the best skillful and loyal human resources for these types of employments: analyst, architect, civil-engineer, strategic planning, middle management, design engineers, mechanical engineer, and functional manager.

On the other hand, the peripheral jobs and competencies that are most subject to contractual employments or routine-based recruitments are: accountants, administrative positions, graphic designers, software engineers, technical jobs and other architectural services, programmers, maintenance, consultants, customer service agents, lawyers and trainers.

In the first category of employment, the core competencies are valuable and strategic for the firm due to their high potential of creativity and innovation. The employers are ready to keep their most skillful employees, to recruit professionals from rival firms by offering more tempting wages, and to invest in training and development of young engineers and architects for potential leadership positions.

In the second category, tasks may be strategic but the peripheral competencies needed in each position have limited creativity. Skills are generic and transferable, and less subject to training and investment in human capital. Firms are inclined to short-term orientation in this employment mode, focusing on goals, productivity and lower wages, and they are more likely to recruit ready-made skillful labor (even at the cost of high turnover).

In conclusion, the globalization of construction services does not imply that partnerships, alliance and subcontracting reduce the potentialities for promotion, loyalty, and career paths within the same company. It only shows that firms are facing emerging challenges in identifying and keeping the most creative professionals with the core competencies. Thus, firms have to find balance between the internal need of investing in professionals in leading positions and the external need of subcontracting, decentralizing, or simply recruiting generic skills and substitutable labor.

Similarly, Higher Education Institutions (HEI) in Lebanon have to evolve in a more globalized educational environment. However, the educational reforms in the private universities are slow, with the main target being the accreditation requirements rather than the establishment of a long-term cooperation with local and regional market.

4.2   The Skill-Supply in Lebanon: The Education Reforms Needed

To analyze the skill-supply in the Lebanese construction sector, one should understand that the educational background of all professionals is regulated by the Order of architects and engineers and by major universities. HEI follow a 5 years program that combines Basic sciences, sciences of engineering, and specific subjects related to each specialty and sub-discipline. The programs are highly focused on analytical skills (such as applied math and sciences) and on modeling and simulating complex systems. The curricula emphasize on different basic creative skills, such as "generating new ideas" or new products and processes, "creating innovative concepts" and original works and projects, and "identifying new trends".

When studying the specialized knowledge and skills in architecture and engineering programs, the competences framework may be divided into 5 major components:

- Analytical knowledge and skills such as calculation, simulation and modeling.
- Factual knowledge and skills that are required from architects and engineers for problem-solving.
- Creative skills.

– Computing skills.
– Interpersonal and intercultural communication skills and attitudes.

The first two components are related to hard-skills and basic knowledge; thus they refer to standardized skills that are common to most graduates, leaving little room for uniqueness, originality and personal talent. These skills and knowledge have a more quantitative nature and they are more subject to objective assessment by teachers, instructors, and supervisors.

The last three components are largely qualitative, and they are marginalized in the curricula (integrated into few courses and associated with other teaching objectives, receiving a small part in the teaching credits). These generic and soft components of the curricula are paradoxically the most needed in work environment and our study shows that the demand for such skills and interpersonal attitudes shapes the future of the profession. This skill mismatch needs to be addressed if freshly graduates are intended to succeed in their transition to the labor market and get promoted to more creative and less routine jobs. Firms are seeking more proactive and creative graduates for leadership positions and tasks, and their limited (and oversaturated) need for analytical knowledge and skills is merely a matter of peripheral work that could be outsourced or easily substituted.

Unfortunately, HEI in Lebanon so far have only been successful in terms of forming standardized engineers and architects with solid analytical and factual knowledge, considering these soft skills as merely supplements, as revealed by the study of employers' perceptions. Table 5 shows for example that potential employers consider that freshly graduate engineers and architects lack soft skills (such as administrative skills) more than analytical skills (Table 6).

Table 6. - Ranking of lacking skills of freshly graduates – PACOME, 2015

Insufficient knowledge of markets needs	76%
Lack of administrative skills	65%
Insufficient training	59%
Insufficient knowledge of rules	58%
Lack of analytical skills	41%
Lack of language mastering	41%
Lack of presentation skills	35%

HEI Should Integrate More Creative Design in Their Curricula and Emphasize More on Engineering Creativity. Indeed, 88% of the respondents in our survey considered design as the main task required from an engineer or an architect in Lebanon. Creativity is the most important soft skill required from an architect (73% of the respondents). The evolutionary design and the formal design evaluation methods are for example still lacking in the Lebanese curricula, while perceived to be core skills of the profession.

New competence frameworks should integrate routine design, with design engineering, inventive design and inventive engineering.

HEI also need to revise their programs, in order to integrate computing and programming fundamentals, as well as the fundamental conceptual understanding of these creative tools and software. Students will be much better prepared for future creative jobs by learning programming and mastering other computer graphic software for 3D animation and design.

Finally, with the ongoing globalization of construction services, young engineers and architects need a better understanding of the global dimension of their actions. With the uprising integration with the Arab Labor market and the Gulf trend, their programs should emphasize more on the common cultural values and intercultural communication skills.

5   Conclusion and Recommendations

While this qualitative research suggests several noteworthy points, more research is needed due to the lack of quantitative studies in Lebanon (data on recruitment needs, frictional unemployment, structural unemployment, average wages, etc.).

Yet, the global, regional and local need for better infrastructure will drive demand for more creative and skilled architects and engineers in the coming years. More skilled professionals are required to conceptualize, design, and implement new projects within a rapidly changing social and political environment.

The key attributes that freshly graduates may lack on their first day of work are team building, strategic thinking, effective communication skills, ethics and respect for other collaborators from different cultural backgrounds. This research indicates that these basic skill-shortages call for a change in the education system that should be enhanced to include information, communication and language competencies. The fast growth in demand over the medium term means that HEI need to form more competent architects and engineers, but this also induces more needs for technical jobs in the construction sectors, especially, IT specialists, and specialist managers adapted to a more flexible and fluid organization of the work place. This requires more vocational and educational training, and accordingly, curricula should not only focus on analytical and technical skills, but also on communication, business management and marketing, in order to meet the employers' needs. A serious reform of educational programs should incorporate these new skills and competencies, such as creative thinking, creative design and creative engineering.

Further, firms need to implement a proper wage policy, and a reward and incentive system, in order to achieve a good balance between investments in training in human capital and the employment policy for peripheral competences and labor. Therefore, our three recommendations are:

1) On the local level, firms and Higher Education Institutions need to strengthen cooperation, for them to compete and survive in a highly integrated regional and global market.

2) Since firms are shifting from a vertical organization to a more horizontal, fluid division of tasks, skills that are largely needed in the future are those that are most likely to reinforce coordination and trust among teams within the internal components of the organization, and to deepen the cooperation with partners' networks.

3) HEI should incorporate in their curricula more creative courses and tools, such as creative design, computing and programming fundamentals, as well as courses related to globalization, with a specific emphasis on intercultural communication skills and attitudes.

References

Arciszewski, T.: Civil engineering crisis. Leadersh. Manag. Eng. **6**(1), 26–30 (2006)

Arciszewski, T.: Change demands renaissance in civil engineering education. Struct. Environ. **3**(4), 5–13 (2011)

Baird, T., Szczygiel, N.: Sociology of professions: the evolution of landscape architecture in the United States. Landscape Rev. **12**(1), 3–25 (2007)

Borrego, M., Henderson, C.: Increasing the use of evidence-based teaching in STEM higher education: a comparison of eight change strategies. J. Eng. Educ. **103**(2), 220–252 (2014)

CAS: National Accounts (2004–2015)

Ejohwomu, O., Proverb D., Olomalaiye P.: A conceptual demand led model founded on theories of labor market mismatch for the construction and building services industry. In: Khosrowshahi, F. (ed.) 21st Annual ARCOM Conference, SOAS London, University of London, vol. 1, pp. 53–62 (2005)

Horns, K., Jenkins, R.: Is the profession of civil engineering becoming a commodity? You should know the answer. Leadersh. Manag. Eng. **11**(1), 40–44 (2001)

Jacobides, M.: The architecture and design of organizational capabilities. Ind. Corp. Change **15**(1), 151–171 (2006)

Kenner, T., Isaak, M.: Leadership development in a civil-engineering culture: finding the balance-point between experience and experiment. Leadersh. Manag. Eng. **4**(3), 105–109 (2004)

Lepak, D., Snell, S.: Examining the human resources architecture: the relationships among human capital, employment, and human resources configurations. J. Manag. **28**(4), 517–543 (2001)

Lucas, J.: Identifying learning objectives by seeking a balance between student and industry expectations for technology exposure in construction education. J. Prof. Issues Eng. Educ. Pract. **143**(3), 05016013 (2016)

Pavlicic, N., Perazic, M., Duric-Jocic, D., Knezevic, M.: Engineering education in the field of civil engineering. J. Appl. Eng. Sci. **12**(1), 11–18 (2014)

Rocha Brito, C.: Engineering education for the improvement of practice: preparing for labor market. Am. Soc. Eng. Educ. AC 2012–4597, 10 p. (2012)

Wu, P., Feng, Y., Pienaar, J., Zhong, Y.: Educational attainment and job requirements: exploring the gaps for construction graduates in Australia from an industry point of view. J. Prof. Issues Eng. Educ. Pract. **141**(4), 06015001 (2015)

Data Analytics

Clustering of Social Media Data and Marketing Decisions

Teissir Benslama(✉) and Rim Jallouli

University of Manouba, Manouba, Tunisia
teissirbenslama@yahoo.fr, rimjallouli@esen-manouba.org

Abstract. The technological revolution and the appearance of Social Media have made it possible to generate large volumes of heterogeneous data called Big Data. Today, Big Data Analytics plays a very important role for businesses in making marketing decisions. Social Media Data represents a large part of Big Data and are characterized by complex and unstructured formats, which makes their analysis a difficult task. The challenge for researchers and decision-makers is to find a path to facilitate the analysis of these huge data in order to extract relevant information and to improve marketing decisions and strategies. In this context, previous research proposed several methods and techniques such as Data Mining, visualization and machine learning. Data Mining techniques are among the most widely used techniques and include Clustering techniques. Clustering provides a wide range of techniques that classify unstructured data and detect useful knowledge from large data sets. In this regard, numerous articles on analyzing Social Media Data using Clustering have been published and there has been a rapid increase in the number of publications in the areas of Social Media Data and marketing, in which several Clustering methods have been proposed. Despite this increase, there is a lack of articles organizing these publications according to Clustering techniques and added value. The aim of this paper is to answer the following questions: What are the techniques for aggregating Social Media Data? What are the marketing decisions generated by Social Media Data Clustering? Thus, it will be useful to present a review and a classification of research articles on Social Media Analysis in the field of marketing using Clustering to provide an overview to researchers and managers looking to use these techniques.

Keywords: Social media · Data analytics · Big data · Marketing decisions · Clustering

1 Introduction: Evolution of Social Media

Human beings have always been in need for communication. The means of communication, developed over the years (Telegraph 1792, telephone 1890, radio 1891) to meet this need, have brought about change and acceleration of information. The first seed of what can be considered as social network has been planted in 1971 by Ray Tomlinson: the first e-mail was exchanged between two computers. In 1978, the first versions of Internet

© Springer Nature Switzerland AG 2020
M. A. Bach Tobji et al. (Eds.): ICDEc 2020, LNBIP 395, pp. 53–65, 2020.
https://doi.org/10.1007/978-3-030-64642-4_5

browsers were distributed on the Usenet platform. Years later, in the 21st century, several important Social Media were developed, involving a technological revolution that has affected all disciplines: Trip Advisor (2000), Wikipedia (2001), LinkedIn (2002), Friendster (2002), MySpace (2003) to name just a few. According to Gilbert and Karahalios (2009) "Social Media is a set of Internet-based applications that is based on the idea of Web 2.0". In 2004, one of the biggest Social Networks was created: Facebook. Facebook opened to the public in 2006 and it changed the history of Social Networks. Companies also saw the opportunity that this new service would offer them in terms of publicity and getting closer to the public. This Social Network is recognized as one of the most important Social Networks of all times. Between 2005 and 2006 YouTube was born and Twitter was created as an internal messaging system to provide news. In recent years, more and more Social Networks have appeared: Tumblr (2007): a microblogging platform, Pinterest (2011): a platform for sharing images between users, and Snapchat (2013): an ephemeral personal agenda with photos and videos. Today millions of people and businesses use these Social Networks which are dynamic, widely spread and form a huge network. Every second, millions of Social Media users spontaneously generate a huge amount of data, called Big Data, that is conveyed, accumulated and unorganized.

Several researchers have analyzed data with different techniques to extract useful information and support marketing decisions, including Clustering techniques. In recent years the number of articles and publications on Social Media Data Analysis for marketing purposes using Clustering has increased significantly (Bello-Orgaz et al. 2020; Jisun et al. 2017; Ianni et al. 2019; Zhang et al. 2017), so there is a need to study which marketing benefits are driven by these techniques in the analysis of Social Media Big Data.

The objective of this research is to measure, firstly, the impact, in a general way, of the Social Media Data Analysis on the marketing decisions according to the existing literature, secondly, to propose a classification of selected recent publications, which used Clustering to analyze Social Media Data for marketing decision making. The classification of these selected papers is based on the following criteria: the type of Social Media, the Clustering technique and the marketing decision orientation. The aim is to present a clear and structured overview of previous results to researchers and professionals aiming to use these techniques.

This paper is organized as follows: In the following section we briefly present the evolution and characteristics of Social Media Big Data as well as the marketing decisions retained from the analysis of these data. The third section introduces the used techniques for analyzing Social Media Big Data. The fourth section, firstly defines Clustering techniques and their impact on marketing decision making when used as a Social Media Data Analysis tool, then presents the methodology of this study. The fifth section explains the main results. Finally, Section six highlights the main points, conclusions and suggestions for future research and draws attention to research limitations.

2   Social Media Data: Evolution and Marketing Decisions

Easy and free access to Social Media has pushed users to publish content easily and without limits. Today more than a trillion users use Social Media to generate voluminous and unstructured data called Big Data or Social Big Data every second in an overwhelming way. Social Media is the most important source of Big Data.

2.1   Evolution of Data Generated by Social Media: Big Social Media Data

"Social Media contents, such as tweets, comments, posts, and reviews, have contributed to the creation of Big Data extensively from any platform providers or different websites" (Lyu and Kim 2016). These data are generated by Social Media in an exponential way and are known by their heterogeneity. In order to organize and define the characteristics of Big Data, researchers have classified these data into three types: "Structured data that constitute only 5% of all existing data" (Cukier 2010), unstructured data (videos, audio, documents, images, comments, likes, tags, tweets, etc.) that require machine analysis, and semi-structured data (graphics, text, etc.).

However, there is not yet a uniform definition of Big Data, some definitions focus on what Big Data is, while others have tried to answer what it does. When we say «Big Data» certainly, size is the first characteristic that comes to mind, however, other characteristics of Big Data have emerged, with the appearance of the famous definition proposed by Laney et al. (2001), the definition of 3V, which characterizes Big Data by volume, variety and velocity. Laney suggested that these 3V are the three dimensions of Big Data challenges. Saggi and Jain (2018) added four more characteristics of Big Data (7V): value, veracity, variability and valence.

Simply extracting these heterogeneous and complex data from Social Media makes no sense, the only thing that gives value and makes these data very useful is analytics. Indeed, Social Media Big Data Analysis brings significant benefits to companies and practitioners in many disciplines for good decision making, including marketing. This enormous amount of data has provided many opportunities and challenges to analyze it. "The most common applications of Big Data for Social Media are trend discovery, Social Media Analytics, sentiment analysis, and opinion mining. For instance, Social Media assist organizations to obtain customers' feedback regarding their products, which can be used to modify decisions and to obtain value out of their business" (Katal and Wazid 2013; Wu et al. 2014).

2.2   Social Media Data and Marketing Decision Making

In the desire to become useful, voluminous data are collected today from Social Media by companies and organizations. This poses the challenge for marketers to analyze this huge amount of data to support marketing decisions, for value creation, and to extract necessary and valuable knowledge. Currently marketing can be considered as the promising field for experimenting with Social Media Big Data Analytics applications, in order to produce benefits for the performance of marketing decisions. The use of Social Media Data for marketing decision making has recently attracted the attention of both practitioners and researchers. Marquez et al. (2019) proposed a Big Data Framework for analyzing Social Media content in order to improve the decision-making process in companies and presented a case study for analyzing the opinions of Internet users on tourism services (hotels and resorts). In order to optimize hotels' decision-making and investments in societal strategic orientation, Koubaa and Jallouli (2019) published a netnographic study of guests' comments concerning hotels of an international chain from a Trip Advisor platform.

Van Dieijen et al. (2019) used a Big Data Analysis of volatility Spillovers of brands across Social Media and stock markets. The results of two studies used on different markets show that specific marketing events lead to the volatility of created content and high volatility spillovers between user-generated content rates on Social Media and stock returns. Rathore and Ilavarasan (2020) compared user emotions towards three new products before and after launch (a pizza, a car, and a smartphone) based on an analysis of 302,632 tweets mentioning the three new products before and after the launch. The results show that consumers' emotions before and after the launch are not the same.

In the current context of rapid growing digital economy driven by emerging technologies and business innovation, significant efforts have been devoted by researchers in the field of marketing and Data sciences to design and propose new techniques and methods that enable the marketing decision makers to analyze data available on Social Media (Bach Tobji et al. 2018; Jallouli et al. 2016, 2017, 2019).

3   Techniques Used for Analyzing Social Media Data

The problem facing companies today is the lack of capabilities to analyze the information collected by Social Media tools Kaabi and Jallouli (2019). Therefore, to effectively analyze Social Media Data, there is an urgent demand for new techniques and analytical methods to process these massive and complex data. In this context, several studies and articles focus on Social Media Big Data Analysis techniques (Ghani et al. 2018; Kim and Hastak 2018; Stieglitz et al. 2018). These techniques provide insights to managers and marketing decision makers. Galetsi et al. (2020) indicated that the main techniques of Big Data Analysis are: optimization, Machine learning, modeling, simulation, visualization, data mining, text mining, web mining, forecasting, and statistics. Rehman et al. (2016) classified Big Data Analysis techniques into machine learning (Supervised learning, unsupervised learning, semi-supervised learning, and deep learning), statistical methods (Descriptive Statistics, and Inferential Statistics) and Data Mining techniques (Classification, and association rules mining).

Lin et al. (2018) used a generalized linear mixed model to analyze the impact of brand engagement on advertising performance for Chinese hotels via Social Media platforms. Liu et al. (2019) used analysis of data extracted from Twitter with natural language processing to examine the impact of luxury brand Social Media marketing on customer engagement. Jabbar et al. (2019) applied real-time Big Data processing on Social Media Data to have instantaneous marketing decisions for B2B industrial marketing organizations. Kirilenko et al (2019) studied tourist attractions Clustering challenges (using data available on TripAdvisor website), according to tourist interests. The study used network analysis, spatial analysis and geo-visualizations to process data. The three clusters (Entertainment, Heritage and Nature) were detected in each domestic market. Rathore and Ilavarasan (2020) developed an approach based on sentiment analysis in their study.

Data Mining techniques are widely used by researchers in Social Media Data Analysis. This paper focuses on research that uses Clustering as an analysis technique given the growing number of articles using this technique in the fields of Social Media Analysis and marketing.

4 Social Media Data Clustering and Marketing Decisions: Research Methodology

Misirlis and Viachopoulou (2018) present a literature review map and a classification of research articles on Social Media metrics and analysis in the marketing field. Several methods and techniques were used: Artificial intelligence, Data Mining and visualization. This paper adopts the same methodology but we will take, as a criterion, a single Data Mining technique: "Clustering". The selection of Clustering is argued by the lack of previous research classifying articles on Social Media according to this technique.

Clustering or unsupervised classification is a Data Mining technique that has proven its effectiveness in data analysis. Clustering aims to divide a set of data into different homogeneous groups, in the sense that the data of each subset share common characteristics, which most often correspond to proximity criteria that are defined by introducing measures and distance classes between objects. Bello-orgaz et al. (2020) define Clustering as follows:

"Clustering can be described as a blind search on a collection of unlabeled data, where elements with similar features are grouped in sets. Elements included in the same cluster should be similar, and elements included in different clusters should be dissimilar. For this reason, it is necessary to define a similarity measure from which this type of algorithm can assign groupings."

Clustering can be applied to all types of data even Big Data such as Social Media Data. Social Media massive and complex amount of data can become greatly reduced after the application of Clustering as its essential role is to homogenize heterogeneous data sets. There are multiple Clustering techniques that have different advantages and can be used together to analyze data, for example: K-means, Fuzzy C-means, Subtractive Clustering, Hierarchical Clustering and PAM Clustering. These techniques are simple to apply, cheap, and effective. Given these advantages, many researchers and practitioners have applied Clustering techniques in Social Media Data Analysis in several fields, such as logistics, finance and marketing.

Many marketers have applied Clustering and processed the Social Media Data extracted by companies to obtain valuable solutions and improvements for marketing strategies and decision-making. The purpose of this study is to present a cartography and classification of research articles using Clustering to analyze Social Media in marketing and decision-making. This paper examines the marketing literature to explain and clarify the measures and the effect of the Clustering' use for Social Media Data in marketing.

The methodology in this paper is guided by the three stages of a content analysis as stated by Bardin (2009): The pre-analysis, the exploitation of the material and finally the treatment, inference and interpretation of results.

Therefore, this study pursues the following protocol: We started with a simple search on two online scientific databases: Science Direct and Google scholar. Science direct provides access to 4 208 Elsevier journals covering a range of disciplines, from the theoretical to the applied research including computer and social sciences. "Google Scholar is increasingly used as a bibliometric tool and competes with Thomson Reuters' Web of Science and Elsevier's Scopus" Moed et al. (2016). The search is done using with each search combinations between the following keywords: "Marketing", "Social Media" and "Clustering". To obtain recent articles, we filtered the results by limiting the search between the year 2016 and 2020. Then to restrict the search to the collection of articles, Encyclopedia and book chapters were excluded from our search. In a second step, to get results more related to the searched keywords we studied only the first 50 articles of each search, and from these results we were interested only in articles with titles related to the field of marketing and decision'making. The search process yielded 33 articles. Then, we carefully studied and revised these 33 articles and we rejected articles with content not compatible with our field of research. The research has thus filtered 20 papers from 2016 to 2020. Only twenty articles were kept to test the effectiveness and efficacy of this method as we intended to apply the same method to a larger number of articles if given valuable results.

After performing a floating reading, we carried out a thematic analysis and established a thematic content analysis grid. According to Robert and Bouillaguet (1997), the purpose of this step is to apply to the corpus of data, treatments allowing access to each thematic idea (keyword or expression or index) referring to one of the following categories: Social Media platform, field of study, type of analysis/method, Clustering technique, marketing objectives and added value. These categories are inspired from metrics used by Misirlis and Viachopoulou (2018). The corpus exploitation was done manually without using software. The results of this research are presented in five tables. Each table classifies the studied papers according to one retained category.

5 Main Results and Discussions

The classification of the articles by the year of publication as shown in Table 1 gives the following frequencies: 3 articles for 2016 and 2018, 4 articles for 2019 and 2020, and 6 articles for 2017. These results explain that since 2016 researchers have continuously used Clustering techniques in the analysis of Social Media Data for marketing and decision-making.

Table 1. The articles 'number per year.

Year	Frequency
2016	3
2017	6
2018	3
2019	4
2020	4

Regarding the platforms used by researchers in the 20 studied articles, the results show the distribution presented in Table 2: 13 articles used twitter, 4 used Facebook, 2 articles studied data using all the Social Media platforms at the same time and only one article for each of E- mail, YouTube, and Linkedin. These results are not surprising since Twitter and Facebook are the most used Social Media platforms in the world, (Misirlis and Viachopoulou 2018; Terzi et al. 2019). Moreover, the extraction and access to data from Twitter is less complex than the other platforms. These results show that there is a need to look for future contributions using platforms other than Facebook and Twitter.

Table 2. Social media platforms 'classification used in the 20 articles: an article can belong to several categories.

Social media platform	Frequency
Twitter	13
Facebook	4
Use of all social media at once	2
E-mail	1
Youtube	1
LinkedIn	1

According to Table 3 most of the articles use the K-means technique which represents the simplest and best-known technique among Clustering techniques (9 articles out of 20). Therefore, there is a utility to apply techniques other than K-means in future studies.

Table 3. Classification of clustering techniques used in the 20 studied articles: an article can belong to several categories.

Clustering technique	Frequency
K-means	9
Hierarchical clustering	3
Modularity clustering	1
Temporal clustering	1
Latent semantic clustering	1
Hybrid variable-scale clustering	1
K-means++	1
K-mussels wandering optimization	1
Novel network clustering method	1
Density-based spatial clustering	1
Mean-shift clustering	1
PAM clustering	1
Sequential clustering	1
Metaheuristics clustering	1

According to Table 4 the most used type of analysis with Clustering in the field of Social Media data and marketing is the Sentiment analysis followed by Content analysis, Text mining analysis, and Algorithmic analysis. The use of sentiment analysis with Clustering techniques is therefore convenient and effective in extracting valuable marketing insight. The list of methods adopted in the studied papers provides a good starting point for researchers as a relevant table that guides future methodological orientations. The frequencies shown in Table 4 can also encourage researchers to apply new methods or rarely adopted methods and explore their effectiveness and relevance depending on the context or the selected platform in the research. Indeed, the choice of Clustering techniques depends on the nature of the data, the area of research, and the capabilities of the researchers.

Table 4. Classification of type of analysis used in the 20 studied articles: an article can belong to several categories.

Type of analysis/method	Frequency
Sentiment analysis	4
Content analysis	3
Text mining analysis	2
Algorithmic analysis	2
Classification analysis	1
Descriptive statistical analysis	1
Algorithmic analysis	1
Multi-dimensional scaling	1
Automated persona generation	1
Communications privacy management	1
Sequential analytical model	1
Data processing	1
Novel big data analytical approach	1
Combined analysis model (behavioral/geographical)	1
Hidden Markov model	1
The DeGroot opinion update model	1

Sixteen fields of study were covered in the studied papers and are listed in Table 5: Travel, Consumer perceptions and behavior, Airport marketing and airline industry, Market segmentation, Brands and cosmetic products, Banking, Supply chain management, Education, E-commerce and privacy behaviors, Halal food, Rare events, User generated content and video marketing, Sales and B2B, Rural E-marketing, Marketing analysis and wineries, News media and political marketing.

For each field and for each article the application of Clustering techniques offered net advantages and added values in the marketing area. As indicated in Table 5 the articles

Table 5. Marketing objectives/value added according to field of study and clustering technique

Field of study	Authors	Clustering technique	Marketing Objectives / Value added
Travel	Amaro et al (2016)	K-means	- Understand traveler's use of Social Media to improve online marketing strategies.
	Zhang et al (2017)	Sequential Clustering	- Support travel behavior modeling in a metropolitan area. - Infer longitudinal travel behavior.
	Hasnat and Hasan (2018)	Density-based spatial Clustering K-means, Mean-Shift Clustering	- Understand travel behavior of tourists.
Consumer perceptions & behavior	Pournarakis et al (2016)	Metaheuristics Clustering	- Produce insights for brand awareness and brand meaning.
	Jisun et al (2017)	Hierarchical Clustering	- Provide competitive marketing insights to businesses.
Airport marketing & airline industry	Punel and Ermagun (2018)	Novel network-Clustering method	- Improve customer strategy for airlines to detect passengers interested in the brands.
	Gitto and Mancuso (2019)	K-means	- Ability to identify airports positions in the perception of the customers.
Market segmentation	Ramasubareddy et al (2020)	K-means	- Offer a strategy for businesses struggling to grow in terms of online marketing.
Brands & Cosmetic products	Kang et al (2016)	K-means	- Help to understand customer preferences.
Banking	Afolabi et al (2017)	K-means	- Extract knowledge to help the banking industry.
Supply chain management	Singh et al (2017)	Hierarchical Clustering	- Improve Supply chain management.
Education	Oliveira and Figeira (2017)	K-means	- Measure the efficiency and return on investments from social contents.
E-commerce & Privacy behaviors	Lankton et al (2017)	Hierarchical Clustering	- Enhance privacy management strategies from Facebook users' behaviors.
Halal food	M. Mostafa (2018)	PAM Clustering	- Investigate attitudes towards halal food expressed on SM and halal food consumer's segmentation.
Rare events	Lu et al (2019)	K-means++, k-MWO K-means	- Support organizations by informing them in advance so that they get ready before the event arrives.
User generated content&Video marketing	Wang and Gao (2019)	Hybrid variable-scale Clustering	- Help managers discover potential customers according to users preferences.
Sales & B2B	Yang et al (2019)	Latent Semantic Clustering	- Identify the product demand trends, enrich marketing strategies early.
Rural E-marketing	Alavion and Taghdisi (2020)	K-means	- Model villagers' intention to adopt e-marketing.
Marketing analysis & wineries	Bello-Orgaz et al (2020)	Temporal Clustering	- Evaluate the Social Behaviors of different brands to help with marketing decisions.
News media & Political Marketing	Stamatelatos et al (2020)	Modularity Clustering	- Obtain valuable information about the political affinity of the participating nodes.

related to travel form the highest number. These results show the richness of options and applications of Social Media Analysis with Clustering techniques. Adopting Clustering techniques to analyze Social Media Data and orient marketing decisions is particularly pertinent during the next years for researchers and marketers.

The results displayed in Table 5 show that several Clustering tools were used according to the objective and the nature of the targeted marketing knowledge and decision. K-means technique was largely the most adopted technique among selected papers in the context of travel, industry, banking, cosmetics, education, rare events and rural e-marketing. K-means Clustering helped extract market knowledge by understanding, modeling customer preferences and intentions and offering a strategy for businesses struggling to grow in terms of online marketing.

The Hierarchical Clustering technique was used in three papers to improve marketing strategy with a focus on supply chain management and e-commerce competitive strategy. The Sequential Clustering technique was selected to support the behavior modeling. The hybrid variable-scale Clustering provided relevant results in the context of user generated content and video marketing. Indeed, the added value of using Hybrid variable-scale Clustering was to help managers discover potential customers according to users' marketing preferences. Finally, the Latent Semantic Clustering technique was implemented to identify the product demand trends and enrich marketing B to B strategies in an earlier stage.

6   Conclusion, Future Research and Limitations

This study mainly aimed to highlight the importance of techniques for aggregating Social Media Data and more precisely the Social Media Data Clustering for marketing decision orientation.

This paper analyzed selected 20 papers from 2016 to 2020 and the results provide researchers and managers with a relevant classification of research on Social Media Analysis in the field of marketing using Clustering.

A content analysis of the selected papers shows that Twitter and Facebook are the most studied platforms and the K-means technique is the most used technique among Clustering techniques. This study highlights a list of methods adopted in the studied papers that provides a good starting point for researchers in future works and points the benefit of applying new methods to explore their effectiveness and relevance for several contexts (the platform, the data, the area of research and the capabilities of the researchers). This research highlights a large set of fields that benefited from using Clustering tools such as travel, banking, education, cosmetics, airports and airline industry, food, events, supply chain, new media and political marketing.

Moreover, this study can help managers have a clear idea on the Clustering techniques of Social Media Data Analysis, and choose the appropriate technique that improves the understanding of customers' intentions and preferences, models market trends and guides businesses struggling to grow in terms of online marketing.

The results of this study should benefit researchers and marketers by helping them better understand what has been achieved in the field of Social Media Data Analysis and marketing using Clustering. The findings of this paper invites academics, practitioners and researchers to focus more on a combination of data analysis capabilities and marketing insights. The main drawback of this study is the limited number of the sample of reviewed papers. This article is indeed a preliminary research that outlines relevant results encouraging future investigations with a larger set of published papers using Clustering techniques to marketing strategy purposes. Future works could address the limitation of this study by analyzing a bigger number of papers selected with more focused keywords. The expected output is a broader and more detailed cartography including more metrics such as the nature of the dataset and the analysis capabilities in order to provide an exhaustive classification available to researchers and practitioners. Another recommendation would be to use software allowing the automatic search of more articles to have an exhaustive literature review of Social Media Data using Clustering for marketing decision guidance.

References

Afolabi, I.T., Ezenwoke, A.A., Ayo, C.K.: Competitive analysis of social media data in the banking industry. Int. J. Internet Mark. Advertising **11**(3), 183 (2017)

Alavion, S.J., Taghdisi.: A. rural e-marketing in Iran; modeling villagers' intention and clustering rural regions. Inf. Process. Agric. (2020). https://doi.org/10.1016/j.inpa.2020.02.008

Amaro, S., Duarte, P., Henriques, C.: Travelers' use of social media: a clustering approach. Ann. Tour. Res. **59**, 1–15 (2016). https://doi.org/10.1016/j.annals.2016.03.007

Bach Tobji, M.A., Jallouli, R., Koubaa, Y., Nijholt, A.: Digital Economy: Emerging Technologies and Business Innovation. LNBIP, vol. 325. Springer, Heidelberg (2018). ISBN 978-3-319-97748-5

Bardin: L'analyse de contenu, 2ème Edition Puf, Quadrillage Manuels (2009)

Bello-Orgaz, G., Mesas, R.M., Zarco, C., Rodriguez, V., Cordón, O., Camacho, D.: Marketing analysis of wineries using social collective behavior from users' temporal activity on Twitter. Inf. Process. Manag. (2020. Article in Press. https://doi.org/10.1016/j.ipm.2020.102220

Cukier, K.: The economist, data everywhere: a special report on managing information. Economist **394**, 3–5 (2010)

Galetsi, P., Katsaliaki, K., Kumar, S.: Big data analytics in health sector: theoretical framework, techniques and prospects. Int. J. Inf. Manag. **50**, 206–216 (2020)

Ghani, N.A., Hamid, S., Targio Hashem, I.A., Ahmed, E.: Big social media data analytics: a survey. Comput. Hum. Behav. (2018). https://doi.org/10.1016/j.chb.2018.08.039

Gilbert, E., Karahalios, K.: Predicting tie strength with social media. In: Proceedings of the 27th International Conference on Human Factors in Computing Systems - CHI 2009 (2009). https://doi.org/10.1145/1518701.1518736

Gitto, S., Mancuso, P.: Brand perceptions of airports using social networks. J. Air Transp. Manag. **75**, 153–163 (2019). https://doi.org/10.1016/j.jairtraman.2019.01.010

Hasnat, M.M., Hasan, S.: Identifying tourists and analyzing spatial patterns of their destinations from location-based social media data. Transp. Res. Part C: Emerg. Technol. **96**, 38–54 (2018)

Jabbar, A., Akhtar, P., Dani, S.: Real-time big data processing for instantaneous marketing decisions: a problematization approach. Ind. Mark. Manag. (2019). https://doi.org/10.1016/j.ind marman.2019.09.001

Jallouli, R., Bach Tobji, M.A., Bélisle, D., Mellouli, S., Abdallah, F., Osman, I.H.: Digital Economy. Emerging Technologies and Business Innovation. LNBIP, vol. 358. Springer, Heidelberg (2019). ISBN 978-3-030-30873-5

Jallouli, R., Nasraoui, O., Bach Tobji, M.A., Srarfi Tabbane, R., Rhouma, R.: International Conference on Digital Economy, ICDEc 2016, Carthage, Tunisia, 28–30 April 2016. IEEE (2016). ISBN 978-1-5090-2230-4

Jallouli, R., Zaïane, O.R., Bach Tobji, M.A., Srarfi Tabbane, R., Nijholt, A.: Digital Economy: Emerging Technologies and Business Innovation. LNBIP, vol. 290. Springer, Heidelberg (2017). ISBN 978-3-319-62736-6

Jansen, B.J., Zhang, M., Sobel, K., Chowdury, A.: Twitter power: tweets as electronic word of mouth. J. Am. Soc. Inform. Sci. Technol. **60**(11), 2169–2188 (2009)

Jimenez-Marquez, J.L., Gonzalez-Carrasco, I., Lopez-Cuadrado, J.L., Ruiz-Mezcua, B.: Towards a big data framework for analyzing social media content. Int. J. Inf. Manag. **44**, 1–12 (2019)

Jisun, A., Haewoon, K., Bernard, J.J.: Automatic generation of personas using youtube social media data. In: Proceedings of the 50th Hawaii International Conference on System Sciences (2017)

Kaabi, S., Jallouli, R.: Overview of e commerce technologies, data analysis capabilities and marketing knowledge. In: International Conference on Digital Economy ICDEc 2019, 15–18, Beirut, Lebanon, 12 p (2019)

Kang, H.-N., Yong, H.-R., Hwang, H.-S.: Brand clustering based on social big data: a case study. Int. J. Softw. Eng. Appl. **10** (4), 27–36 (2016). https://doi.org/10.14257/ijseia.2016.10.4.03

Katal, A., Wazid, M., Goudar, R.H.: Big data: issues, challenges, tools and good practices. In: 2013 Sixth International Conference on Contemporary Computing (IC3) (2013). https://doi.org/10.1109/ic3.2013.6612229

Kim, J., Hastak, M.: Social network analysis. Int. J. Inf. Manag. J. Inf. Prof. **38**(1), 86–96 (2018)

Kirilenko, A.P., Stepchenkova, S.O., Hernandez, J.M.: Comparative clustering of destination attractions for different origin markets with network and spatial analyses of online reviews. Tour. Manag. **72**, 400–410 (2019)

Koubaa, H., Jallouli, R.: Social networks and societal strategic orientation in the hotel sector: netnographic study. In: International Conference on Digital Economy (2019)

Laney, D.: 3D Data Management: Controlling Data Volume, Velocity, and Variety. META group Inc., 2001 (2013). http://blogs.gartner.com/doug-laney/files/2012/01/ad949-3D-Data-Management-Controlling-Data-Volume-Velocity-and-Variety.pdf

Lankton, N.K., McKnight, D.H., Tripp, J.F.: Facebook privacy management strategies: a cluster analysis of user privacy behaviors. Comput. Hum. Behav. **76**, 149–163 (2017). https://doi.org/10.1016/j.chb.2017.07.015

Ianni, M., Masciari, E., Mazzeo, G.M., Mezzanzanica, M., Zaniolo, C.: Fast and effective big data exploration by clustering. Future Gener. Comput. Syst. **102**, 84–94 (2019). https://doi.org/10.1016/j.future.2019.07.077

Lin, S., Yang, S., Ma, M., Huang, J.: Value co-creation on social media. Int. J. Contemp. Hospital. Manag. **30**(4), 2153–2174 (2018). https://doi.org/10.1108/ijchm-08-2016-0484

Liu, X., Shin, H., Burns, A.C.: Examining the impact of luxury brand's social media marketing on customer engagement: using big data analytics and natural language processing. J. Bus. Res. (2019). https://doi.org/10.1016/j.jbusres.2019.04.042

Lu, X.S., Zhou, M., Qi, L., Liu, H.: Clustering-algorithm-based rare-event evolution analysis via social media data. IEEE Trans. Comput. Soc. Syst. 1–10 (2019). https://doi.org/10.1109/tcss.2019.2898774

Lyu, K., Kim, H.: Sentiment analysis using word polarity of social media. Wirel. Pers. Commun. **89**(3), 941–958 (2016)

Misirlis, N., Vlachopoulou, M.: Social media metrics and analytics in marketing – S3M: a mapping literature review. Int. J. Inf. Manag. **38**(1), 270–276 (2018). https://doi.org/10.1016/j.ijinfomgt.2017.10.005

Moed, H.F., Bar-Ilan, J., Halevi, G.: A new methodology for comparing Google Scholar and Scopus. J. Informetrics **10**(2), 533–551 (2016). https://doi.org/10.1016/j.joi.2016.04.017

Mostapha, M.: Clustering halal food consumers: a Twitter sentiment analysis. Int. J. Mark. Res. 1–18 (2018). https://doi.org/10.1177/1470785318771451

Oliveira, L., Figueira, A.: Improving the benchmarking of social media content strategies using clustering and KPI. Procedia Comput. Sci. **121**, 826–834 (2017)

Pournarakis, D.E., Sotiropoulos, D.N., Giaglis, G.M.: A computational model for mining consumer perceptions in social media. Decis. Support Syst. **93**, 98–110 (2017). https://doi.org/10.1016/j.dss.2016.09.018

Punel, A., Ermagun, A.: Using Twitter network to detect market segments in the airline industry. J. Air Transp. Manag. **73**, 67–76 (2018). https://doi.org/10.1016/j.jairtraman.2018.08.004

Ramasubbareddy, S., Srinivas, T.A.S., Govinda, K., Manivannan, S.S.: Comparative study of clustering techniques in market segmentation. Innov. Comput. Sci. Eng. **103**, 117–125 (2020)

Rathore, A.K., Ilavarasan, P.V.: Pre- and post-launch emotions in new product development: Insights from twitter analytics of three products. Int. J. Inf. Manag. **50**, 111–127 (2020)

ur Rehman, M.H., Chang, V., Batool, A., Wah, T.Y.: Big data reduction framework for value creation in sustainable enterprises. Int. J. Inf. Manag. **36**(6), 917–928 (2016)

Robert, A.D., Bouillaguet, A.: L'analyse de contenu. Presses universitaires de France (1997).

Sautter, E.T., Leisen, B.: Managing stakeholders a tourism planning model (1999)

Saggi, M.K., Jain, S.: A survey towards an integration of big data analytics to big insights for value-creation. Inf. Process. Manag. **54**(5), 758–790 (2018). https://doi.org/10.1016/j.ipm.2018.01.010

Singh, A., Shukla, N., Mishra, N.: Social media data analytics to improve supply chain management in food industries. Transp. Res. Part E: Logist. Transp. Rev. **114**, 398–415 (2017)

Stamatelatos, G., Gyftopoulos, S., Drosatos, G., Efraimidis, P.S.: Revealing the political affinity of online entities through their Twitter followers. Inf. Process. Manag. **57**(2) (2020)

Stieglitz, S., Mirbabaie, M., Ross, B., Neuberger, C.: Social media analytics – challenges in topic discovery, data collection, and data preparation. Int. J. Inf. Manag. **39**, 156–168 (2018)

Terzi, B., Bulut, S., Kaya, N.: Factors affecting nursing and midwifery students' attitudes toward social media. Nurse Educ. Pract. **35**, 141–149 (2019). https://doi.org/10.1016/j.nepr.2019.02.012

Van Dieijen, M., Borah, A., Tellis, G.J., Franses, P.H.: Big data analysis of volatility spillovers of brands across social media and stock markets. Ind. Mark. Manag. (2019). https://doi.org/10.1016/j.indmarman.2018.12.006

Wang, A., Gao, X.: Hybrid variable-scale clustering method for social media marketing on user generated instant music video. Tech. Gazette **26**(3) (2019). https://doi.org/10.17559/tv-20190314152108

Wu, X., Zhu, X., Wu, G.-Q., Ding, W.: Data mining with big data. IEEE Trans. Knowl. Data Eng. **26**(1), 97–107 (2014)

Yang, Y., See-To, E.W.K., Papagiannidis, S.: You have not been archiving emails for no reason! Using big data analytics to cluster B2B interest in products and services and link clusters to financial performance. Ind. Mark. Manag. (2019). https://doi.org/10.1016/j.indmarman.2019.01.016

Zhang, Z., He, Q., Zhu, S.: Potentials of using social media to infer the longitudinal travel behavior: a sequential model- based clustering method. Transp. Res. Part C: Emerg. Technol. **85**, 396–414 (2017). https://doi.org/10.1016/j.trc.2017.10.005

The Effect of Big Data Analytics on Firm Decision Making: The Case of the Lebanese Banking Sector

Lina Shouman$^{(\boxtimes)}$ (iD) and Jamal Chehade

Lebanese International University, Beirut, Lebanon
`Lina.shouman@liu.edu.lb`

Abstract. The purpose of this paper is to study the perceived impact of big data analytics, a subset of business analytics, on the decision-making process in the banking sector considering that decisions made using big data analytics uncover unseen innovation opportunities and improve compliance within a more stringent regulatory environment. These large complex data sets also known as big data are used by many organizations to enhance their business operations, address business problems as well as generate new opportunities. Big Data Analytics (BDA) analyze Big Data (BD) to uncover relevant information such as customer preferences, hidden patterns, market trends and unknown associations. Data analytics are frequently instilled among the most effective businesses supporting their decision-making process. This paper discusses the impacts of big data analytics on the decision-making process in financial organizations knowing that these analytics can create a constant flow of possible new insights. A qualitative approach has been adopted by conducting in depth interviews with executives in the financial sector to examine how data analytics impact every stage of the decision-making process.

Keywords: Big Data · Big Data Analytics · Decision-making · Financial sector

1 Introduction

The fast development of technology such as the Internet of things, artificial intelligence, big data and cloud computing has interrupted many traditional sectors creating new opportunities; the financial sector is not an exception. Nowadays, organizations' access to information has increased tremendously and organizations are putting tremendous effort into discovering new means to use the accessible data (Harfouche et al. 2019). Big data has had an important effect in many industries and economies around the globe such as healthcare, education, and production and retail. Based on a report by Researchmoz (2014), the financial services sector has a lot more to acquire by leveraging big data; the technology will not only enable financial institutions to exploit the significance of data, but will also allow them to achieve a competitive advantage, operational efficiency, transform obstacles into opportunities and reduce risks in real time.

© Springer Nature Switzerland AG 2020
M. A. Bach Tobji et al. (Eds.): ICDEc 2020, LNBIP 395, pp. 66–75, 2020.
https://doi.org/10.1007/978-3-030-64642-4_6

Several researchers have stated that technologies that are related to BD can be applied in many areas of the banking sector. These areas include retail banking (bank collections, credit cards, private banking), commercial banking (credit risk analysis, customer and sales management, middle market loans), capital markets (negotiation and sales, structured finance) and asset management (wealth management, management of capital investments, global asset reporting, analysis of investment deposits) (Lackovic et al. 2016; Mohamad et al. 2015). In this regard, the financial services sector is utilizing BD to transform its business processes, its organizations, and the sector as a whole (Turner et al. 2013). Thus, financial institutions and banks are no longer questioning the advantages of BD today; they are confident that BDA can offer them an important competitive strength (Bedeley 2014).

1.1 Overview of the Lebanese Financial Sector

In so many countries around the world, banks are considered the most significant financial intermediaries and perform an important role in linking savings and investments, Lebanon is one example. The Lebanese economy is service-oriented and is primarily founded on financial services, trade and tourism. Agriculture, industry and services represent 5.1%, 15.9% and 79% of the Gross Domestic Product respectively. The banking sector is considered a major component of the financial sector and subsequently a key pillar of the Lebanese economy. Lebanese banks particularly focus on bringing in foreign funds, which generate a vital source of reserves for the Lebanese Central bank. The Lebanese banking sector is distinguished by its conservative performance, which enabled it to survive the international financial crisis of 2007–2008, as well as past and present local and regional instabilities. The banking sector in Lebanon manifested itself as the most robust in the Middle East and North Africa (MENA) region during turmoil and stayed very desirable to capital inflows as demonstrated by the high growth rate of deposits from the year 2006 till 2011 (Awdeh 2012).

2 Literature Review

Big data is a term used to describe the surge in the volume of data that are challenging to store, process, and analyze using traditional database technologies. Boyd and Crawford (2012) have described BD as "a cultural, technological, and scholarly phenomenon that rests on the interplay of technology, analysis, and mythology that provokes extensive utopian and dystopian rhetoric". While the usage of BD has become usual and extensive and its rising momentum has gained reputation as an important driver of organizations' currently pursued strategies (Cao et al. 2019), knowing whether BD has an impact on anticipated future strategies is nonexistent in the empirical literature. Remarkably, BD has a proactive analytics power and predictive value (Baesens et al. 2016; Tabesh et al. 2019) that can be a guide for future strategies. This study aims to examine whether the use of BD (i.e. the implementation of BD systems into the organization business processes) and BDA has an impact on the decision-making process of enterprises and specifically the banking sector.

2.1 Big Data Analytics

BDA has its roots in the earlier data analysis methodologies using statistical techniques such as regression, factor analysis, etc. It includes data mining from high speed data streams and sensor data to get real time analytics (Chen et al. 2012). It is an interdisciplinary field which uses knowledge of computer science, data science, statistics and mathematical models. It consists of a systematic process of capturing and analyzing business data, developing a statistical model either to explain the phenomenon (Descriptive Analytics), developing a model to predict future outcomes based on variable inputs (Predictive Analytics) or developing a model to optimize or simulate outcomes based on variations in inputs (Prescriptive Analytics). BDA leverages statistical techniques such as regression, factor analysis, multivariate statistics and knowledge of mathematics for developing equations (Dubey and Gunasekaran 2015). LaValle et al. (2011) conducted an exploratory study on BDA and the path from insights to value. They reported that, with an improving technology, there has been an enormous collection of BD and researchers are still in the way for finding the best ways to analyze these data so that they can obtain valuable information. In the present era, researchers and people are not concerned with "what happened" or "why it happened" commonly known as descriptive analytics. But the main issue of concern is to find out the answer to questions like "what is happening in the present" and "what is likely to happen in the future" commonly known as predictive analytics; and "what actions should be taken to find out the optimal results" basically known as prescriptive analytics. Therefore, business analytics can be classified into descriptive, predictive and prescriptive analytics as explained in Fig. 1 below. Predictive Analytics is elaborated with further details in the next section considering its significance for various stakeholders in the society and business.

BDA is not just about the amount of data, but about the variety of information. Russom (2011)'s research discovered that a very small percentage of people is conscious of concepts such as big data analytics, advanced analytics and predictive analytics.

BDA significantly affect business value and company efficiency, resulting in savings, reduced operating costs, communications expenses, increased returns, improved client relationships, and new company plans (Akter et al. 2016).

2.2 Benefits of Using Big Data in Decision Making

With the emergence of BD, the executives' need for information has been altered. The large datasets coming from a variety of sources in structured, semi-structured or unstructured forms provide forms with several ways to tap value from these datasets and make strategic, tactical and operational decisions. When business transaction data are mined for association rules, they provide key insights for decision makers about products bought together or for predicting future demand for certain items. Understanding those patterns helps retailers such as Walmart redesign their shelves and place products together which eventually leads to improved sales (Shaw et al. 2001). Predicting demand for some products helps improve pre-planning in cases of major natural disasters like hurricanes (Shaw et al. 2001). In addition, leading banks, like Bank of America and Wells Fargo, are also using BDA to comprehend some aspects of their clients' relationship that they couldn't previously obtain. They are observing customers' "journeys" through the use

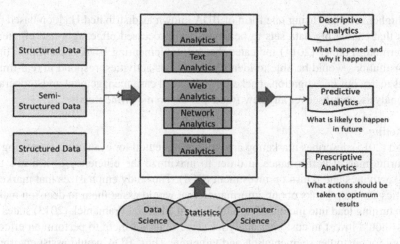

Fig. 1. Framework for descriptive, predictive and prescriptive analytics

of websites, call centers, tellers, and other branch personnel to figure out the directions that customers follow through the bank, and how those directions impact the purchase of specific financial services (Davenport and Dyché 2013). In addition, using BD presents various opportunities of cost and time reduction. Compared to traditional databases, BD techniques are emerging as considerably low-cost solutions. In real-time decision making, BD plays an important role in promoting offers and services to consumers based on their locations. For instance, UPS saves millions of dollars in fuel by gathering and analyzing information from telematics sensors mounted on its 46,000 cars and redesigning car paths using BD (Davenport and Dyché 2013).

2.3 BDA Impact on the Financial Sector

According to The Financial Brand (2014), financial institutions are striving for ways to take advantage of big data. Banks for instance use only an insignificant amount of information to generate insights and improve clients' experiences. The article claims that banks applying BDA have a market share lead of 4% over banks that do not. BDA are used in the following main areas: fraud detection, marketing and risk assessment.

Fraud Detection
Urban (2014) reveals that BDA have become an important element of any approach that can be used for identifying and deterring financial crime. Due to the ever-evolving techniques of attacks that are created by criminals to exploit multi-channel vulnerabilities and compromise information systems, BDA are deemed imperative. BDA have allowed banks to perform large-scale real-time analytics to cope with growing threats.

Kumar (2015) explains that fraud has caused annual losses of more than $1,744 billion, while the banking sector continues to spend millions every year on technology directed at reducing fraud and retaining clients. Kumar suggests a broad fraud detection method that identifies both known and novel fraud cases as they happen in real time, with higher accuracy using one form of BDA known as distributed Hadoop based platforms that enable big data sets to be stored and processed effectively and efficiently. Furthermore, Palmer (2014) indicates that by implementing BDA capabilities, financial institutions would be able to identify suspicious activities, respond in real-time to threats, carry out investigations backed by advanced case management and cooperative data analysis, and discover any new fraud and economic crime trends.

Marketing

Xerago (2015) describes marketing analytics as a method for evaluating, managing and measuring market performance in order to maximize the efficiency of a firm's marketing activities and return on investment (ROI). The study emphasizes that marketing analytics can help banks obtain information that would assist them in decision making and in turning lead into increased profitability. In addition, Pramanick (2013) states that banks should invest in customer analytics since it allows them to perform an effective strategy for customer segmentation and targeting. Thus, BDA would assist enterprises in determining clients' strategies, devising new products and services, and formulating appropriate pricing and marketing techniques. Morabito (2015) declares that marketing automation enabled by BDA will help banks better serve individual customers' needs while maintaining a low cost of marketing expenses. Thus, BDA would enable a personalized customer experience at a good ROI.

Credit Risk Management

Credit risk is the likelihood of loss due to the inability of a borrower to make any type of debt payments. Managing credit risk is the practice of mitigating those losses by comprehending the adequacy of both loan loss reserves and bank capital. According to a study by the financial services association of the European financial services community, country members have started deploying BDA to their credit risk models. BDA techniques help bankers gain a greater insight into the behaviors of their customers by evaluating credit reports, expenditure habits, and repayment rates of loan applicants. Big data tools identify the probability of a customer to default on a loan or continuously fail to meet payment deadlines (Hortonworks Inc. 2013).

3 Methodology

This paper aims to answer the following research question: what is the impact of big data analytics on the decision-making process in the banking sector?

For that purpose, a qualitative approach has been adopted by conducting in depth interviews with executives that work in the financial sector and whose work is related to BD. These face-to-face interviews make it possible to identify the emotions, opinions and feelings of the participants with respect to the subject of research (Bolderston 2012).

The semi-structured interviews were conducted with ten executives from the Lebanese financial sector who are actively using big data analytics for effective subordinates' management and improved decision-making. All ten interviewees work in the private sector and have more than three years of experience with BDA. Some of the executives have implemented the BDA tools at their organizations and hold executive positions such as CTOs, while others manage employees who use BDA and obtain from them relevant information that assists them in the decision making process such as CFOs. The interviewees mentioned that the major disadvantage to working with BD was the data quality issues. Before being able to use big data for analytics, they need to make sure that the information they are receiving is correct, significant and in the right format for analysis. Another problematic issue is conforming to government regulations. Much of the information included in banks' big data databases is sensitive and personal, meaning that the bank may need to ensure that it is complying with government requirements when handling and storing the data.

The purpose of the interviews is to examine how BDA directly affect decision-making in organizations and organizational intelligence. The interviews are divided into four categories based on Huber's theory classifications (Huber 1990). The following classifications are: the impacts of using BDA at subunit level, at organizational level, on organizational memory, and on decision-making and organizational intelligence.

The first part of the interview is used to assist the interviewer in better understanding the interviewees, their positions in the organization, the decisions they have to make, and their connection with data analytics. Interviewees were also asked about their organization's strategy about BDA and the departments that are involved in using them. In the second part of the interview, the interviewees are questioned about the impact of BDA on the organization decision-making process using Simon's model (Simon 1977). The model implies that the three stages of the decision-making process as intelligence, design and choice are addressed individually.

The third part of the interview examines how data analytics directly affect decision-making in organizations and organizational intelligence based on Huber's theory (Huber 1990).

As for the fourth part, interview questions are used to evaluate the impact of BDA on organizational memory keeping in mind that BDA is an essential element and factor that directly affects decision-making (Huber 1990). Furthermore, the fifth part of our interview focuses on the impacts of BDA that are produced at the organization's sub-unit level and yet affects multiple elements of Huber (1990) decision-making such as the size and composition of decision-making units, involvement in decision-making and the time consumed on decision-making discussions. Moreover, part six of the interview addresses the impact of BDA at the organizational level and based on Huber's theory. The items being discussed are the centralization of decision-making, the managerial levels engaged in the decision-making process, and the impact of using BDA at all organizational levels in line with Huber hypothesis (Huber 1990).

Finally, the last part of our interview aims at closing the interview, asking for general view on BDA and making sure that the interviewees have freely expressed their views.

3.1 Data Analysis

The researcher applied the Thematic Content Analysis approach, which is the most commonly used analysis method in qualitative research. To complete the analysis, the researcher applied the six phases developed by Braun and Clarke (Braun and Clarke 2006). Analysis can be an ongoing procedure that is carried out simultaneously while collecting data (Creswell 2009).

Getting Familiar with the Data: In preparation for the next evaluation phase, the researcher turned all the collected notes and audio recordings into text. Then the researcher started reading through the text and taking initial notes.

Coding or Labeling: Once the research became familiar with the data, he began the coding process. During this process, the researcher highlighted parts of the text such as sentences, and created labels or "codes" to describe their content. Then, data sorting and analysis was completed using NVivo software, which helped the researcher achieve results that are more vigorous in a shorter period.

Searching for Themes: Next, the researcher examined the created codes and identified patterns among them, and began developing themes. The research combined several codes into a single theme. For instance, the codes uncertain, let's ask the experts and broad explanations were all combined under the "uncertainty" theme.

Reviewing Themes: During this phase, the researcher made sure that the created themes were valuable and represented the collected data accurately. To do so, the researcher returned to the collected data and compared them with the themes. This comparison allowed the researcher to determine if these themes really present the data and to perform the necessary changes to make the themes work better.

Defining and Naming Themes: In the defining themes phase, the researcher communicated exactly what is meant by each theme and determined how it helps in understanding the data. As for the naming part, the researcher created concise and easy to understand name for each theme.

Writing Up: This final phase the researcher merged the analytics with data extracts in order to create a comprehensible narrative that contains quotes from the interviewees. Furthermore, the researcher tried to relate the analysis to the research question and previously reviewed literature.

4 Findings

The study aimed at generating knowledge about the perceived impacts of using data analytics on organizational decision-making. Towards this end, a qualitative study was carried out to answer our research question: what is the influence of Big Data Analytics on decision making?

The research question is tackled by evaluating the perceived impacts of BDA on decision-making using Simon's decision-making model (Simon 1977) and Huber's theory of the impacts of data analytics on organizational design, intelligence and decision-making (Huber 1990).

With regard to the perceived impacts of the use of data analytics in organizations evaluated through the perspective of the Simon model (Simon 1977), our research shows that data analytics promotes the decision-making process intelligence and design stage, but does not affect the choice stage. Using BDA enables organizations to gain a better knowledge of the environment in which they operate, discover new opportunities, identify requirements for decision-making, and stimulate data gathering and sharing in organizations as intelligence-related activities.

Furthermore, BDA assists the organization with the design stage in the creation, assessment, comparison and prioritization of situations. However, BDA does not affect the decision-making choice stage and the decision-makers always have the final word when making the final decision.

Hereby, in accordance with Huber's theory (Huber 1990), the research shows that at the subunit level the use of BDA affect decision-making involvement in terms of reducing the number of individuals who are required to make decisions. In addition, it can be stated that BDA generally affects the size and heterogeneity of decision divisions in the organization by decreasing the number of individuals that make up the decision units and the frequency of meetings. At the organizational level, the use of data analytics decentralizes the decision-making process allowing individuals at lower hierarchical levels to produce and authorize choices at a satisfactory rate for the organization. In addition, it is considered that the use of data analytics reduces the amount of organizational levels engaged in authorizing activities linked to making a certain choice. As for BDA effects on organizational memory, the use of data analytics triggers more frequent growth and use of organizational memory components, thus affecting the decision-making process by offering accurate, current and quality information. In addition, using data analytics results in a more efficient environmental scanning to identify threats and opportunities faster and more accurately. Moreover, data analytics enhances the quality, relevance, and accessibility of organizational intelligence used in decision making. BDA increases decision-making speed in terms of reducing the time required for a decision and authorizing the actions to be taken.

Finally, the greatest and most significant perceived impact of BDA in organizations is its contribution to a significant improvement in the performance of the choices taken. Data analytics is regarded as a distinctive resource that provides the organization a competitive advantage by enhancing the quality of their choices. BDA can be best utilized to forecast customer behavior, business trends, deliver products to the right clients at the right timing, attract and retain the most valuable customers, and start a new marketing campaign or participate in new markets.

Although organizations understand how significant data analytics are for their decision-making process, the research shows that decision makers do not solely depend on the results of BDA.

Great value resides in the expertise of the Lebanese decision-makers as well as in the reliance on their intuition due to the absence of historical information. In conclusion, in Lebanese financial institutions, the decision making process does not rely exclusively on data analytics which is however used to complement the process.

5 Recommendation

BDA is the trendiest topic in technology in many industries including the banking sector. By using the right analytical techniques, BDA can help firms enhance their performance (Akter et al. 2016). Data analytics is regarded as playing a significant role in the organizational decision-making process. Organizations are aware today more than ever that

BDA can aid them in forecasting customer behavior and business trends, delivering the right products to the right customers at the right moment, designing the appropriate marketing campaign and participating in new industries. They believe that BDA can enable them in achieving all these goals as well as provide them with a competitive advantage, which is of vital significance to their survival.

Furthermore, BDA assists organizations in the creation, assessment, comparison and prioritization of situations as possible course of action in the decision-making process. However, in this study aimed at generating knowledge about the perceived impacts of using BDA on organizational decision-making, the results show that BDA is not considered to affect the decision-making process choice stage, and the decision-makers are always the ones with the final word when choosing the option as a final decision. Financial organizations in Lebanon should apply BDA to improve decision-making, uncover unseen innovation opportunities and improve compliance within a more stringent regulatory environment.

Future research should include more organizations in the study so that many issues can be investigated and more commonality in the findings can be verified. Comparisons can be made between organizations, and the results and conclusions can be generalized with more confidence and reliability. The sample organizations can also be extended to organizations operating in various industries, such as manufacturing, services, and health sectors to say the least. Different industries may have their own unique big data issues. Finally, other methods of data collection, such as survey questionnaires, can be employed to attract a larger number of respondents, particularly those directly responsible for BDA.

References

Akter, S., Wamba, S.F., Gunasekaran, A., Dubey, R., Childe, S.J.: How to improve firm performance using big data analytics capability and business strategy alignment? Int. J. Prod. Econ. **182**, 113–131 (2016)

Awdeh, A.: Banking sector development and economic growth in Lebanon. Int. Res. J. Finance Econ. **100**(1), 54–62 (2012)

Baesens, B., Bapna, R., Marsden, J.R., Vanthienen, H., Zhao, J.L.: Transformational issues of big data and analytics in networked business. MIS Q. **40**(4), 807–818 (2016)

Bedeley, R.T., Iyer, L.S.: Big data opportunities and challenges: the case of banking industry. In: Proceedings of the Southern Association for Information Systems Conference, vol. 1, pp. 1–6, March 2014

Bolderston, A.: Conducting a research interview. J. Med. Imaging Radiat. Sci. **43**(1), 66–76 (2012)

Boyd, D., Crawford, K.: Critical questions for big data: provocations for a cultural, technological, and scholarly phenomenon. Inf. Commun. Soc. **15**(5), 662–679 (2012)

Braun, V., Clarke, V.: Using thematic analysis in psychology. Qual. Res. Psychol. **3**(2), 77–101 (2006). https://doi.org/10.1191/1478088706qp063oa

Cao, G., Duan, Y., El Banna, A.: A dynamic capability view of marketing analytics: evidence from UK firms. Ind. Mark. Manag. **76**, 72–83 (2019)

Chen, H., Chiang, R.H., Storey, V.C.: Business intelligence and analytics: from big data to big impact. MIS Q. **36**(4), 1165–1188 (2012)

Creswell, J.W.: Research Design: Qualitative, Quantitative, and Mixed Methods Approaches. Sage Publications Inc, Thousand Oaks (2009)

Davenport, T.H., Dyché, J.: Big data in big companies. Int. Inst. Anal. (2013)

Dubey, R., Gunasekaran, A., Childe, S.J., Wamba, S.F., Papadopoulos, T.: The impact of big data on world-class sustainable manufacturing. Int. J. Adv. Manuf. Technol. **84**(1–4), 1–15 (2015)

Harfouche, A.L., et al.: Accelerating climate resilient plant breeding by applying next-generation artificial intelligence. Trends Biotechnol. (2019)

Hortonworks Inc.: Banks looking for big data solutions to address credit risks (2013). http://hortonworks.com/big-data-insights/banks-looking-for-big-data-solutions-to-address-credit-risks. Accessed 20 July 2019

Huber, G.P.: A theory of the effects of advanced information technologies on organizational design, intelligence, and decision making. Acad. Manag. Rev. **15**(1), 47–71 (1990)

Kumar, N.: New age fraud analytics: machine learning on Hadoop, 13 March 2015. https://www.mapr.com/blog/new-age-fraud-analytics-machine-learning-hadoop#.Va1Iv_kQjm4. Accessed 22 July 2019

Lackovic, D.I., Kovska, V., Lakovic, V.Z.: Framework for big data usage. In: Risk Management Process in Banking Institutions. Central European Conference on Information and Intelligent System, pp. 49–54 (2016)

LaValle, S., Lesser, E., Shockley, R., Hopkins, M.S., Kruschwitz, N.: Big data, analytics and the path from insights to value. MIT Sloan Manag. Rev. **52**(2), 21–32 (2011)

Mohamad, S.H., Rashila, R., Marwan, Y.I.M., Azzam, I.R.: Reputation risk and its impact on the Islamic banks: case of the Murabaha. Int. J. Econ. Financ. Issues **5**(4), 854–859 (2015)

Morabito, V.: Big Data and Analytics: Strategic and Organizational Impacts. Springer, Heidelberg (2015). https://doi.org/10.1007/978-3-319-10665-6

Palmer, B.: Fighting fraud in banking with big data and analytics (2014). http://www.ibmbigdatahub.com/blog/fighting-fraud-banking-big-data-and-analytics. Accessed 22 July 2019

Pramanick, S.: Analytics in Banking Services (2013). http://www.ibmbigdatahub.com/blog/analytics-banking-services. Accessed 21 July 2019

Researchmoz: Big data in financial services industry: market trends, challenges, and prospects 2013–2018 (2014)

Russom, P.: Big data analytics. TDWI Best Practices Report, Fourth Quarter, pp. 1–35 (2011)

SAS: Credit risk management: what it is and why it matters (no date). http://www.sas.com/en_us/insights/risk-fraud/credit-risk-management.html

Shaw, M.J., Subramaniam, C., Tan, G.W., Welge, M.E.: Knowledge management and data mining for marketing. Decis. Support Syst. **31**(1), 127–137 (2001)

Simon, H.: The New Science of Management Decision. Prentice Hall, Englewood Cliffs (1977)

Snieder, R., Larner, K.: The art of being a scientist: a guide for graduate students and their mentors (2009). https://doi.org/10.1017/cbo9780511816543. Accessed 17 July 2019

The Financial Brand: Big data: profitability, potential and problems in banking (2014). http://thefinancialbrand.com/38801/big-data-profitability-strategy-analytics-banking. Accessed 22 July 2019

Tabesh, P., Mousavidin, E., Hasani, S.: Implementing big data strategies: a managerial perspective. Bus. Horizons **62**(3) (2019)

Turner, D., Schroeck, M., Shockley, R.: Analytics: the real-world use of big data in financial services. IBM Glob. Bus. Serv. **27** (2013)

Urban, M.: Big data analytics in the fight against financial crime (2014). http://www.bobsguide.com/guide/news/2014/Dec/18/big-data-analytics-in-the-fight-against-financial-crime.html. Accessed 23 July 2019

Xerago: Marketing analytics (2015)

A Trust-Based Clustering Approach for Identifying Grey Sheep Users

Ghassen Bejaoui(✉) and Raouia Ayachi(✉)

Higher Institute of Management of Tunis, Tunis, Tunisia
bejaoui.ghassen@yahoo.fr, raouia.ayachi@gmail.com

Abstract. In the context of recommender systems, users may have particular tastes and very unusual preferences comparing to the others. These users are called Grey Sheep Users. It is difficult to find similar users and relevant recommendations for such kind of users. In this work, we deal with trust values in learning users' behaviours and relations between each other. A trust-based clustering approach is proposed for identifying Grey Sheep Users. The obtained cluster is then exploited to make recommendations for target unusual users.

Keywords: Recommender systems · Collaborative filtering · Grey sheep problem · Trust

1 Introduction

Recommender systems have become extremely important in the last few years due to the digital revolution triggered by the creation of social networks. Collaborative filtering is one of the most successful recommendation techniques [3]. It is based upon the information collected about the different users, most of the time preferences, and assumes that users' preferences are consistent among users. In order to infer the active user's preferences, collaborative filtering relies on the preferences of a community of users [13]. The resources with the highest estimated preferences are then recommended to the active user.

Collaborative filtering suffers from several problems such as sparsity, cold-start, scalability etc. Another problem has also been studied and concerns users whome opinions do not match with any group of users and, therefore, are unable to get benefit from collaborative filtering, it is called the Grey Sheep Users problem (GSU).

Few attempts have been proposed in literature in order to solve the grey sheep problem [4,6,9,14,15]. All GSU identification approaches are based on models, either clustering or matrix factorization. They only rely on similarity between users, or this measure can be insufficient for this special case of users since they have different tastes and preferences comparing to the community. Taking only into account ratings is insufficient because other factors (e.g demographic factors, trust relations, etc.) can make the identification phase of this type of users more

© Springer Nature Switzerland AG 2020
M. A. Bach Tobji et al. (Eds.): ICDEc 2020, LNBIP 395, pp. 76–88, 2020.
https://doi.org/10.1007/978-3-030-64642-4_7

accurate and as a consequence make the recommender systems outputs more pertinent.

So, our idea is to take the advantage of trust relations in web-based systems in order to understand more and more the relation between users and look beyond the classical similarity based methods. Such relations can be very useful in bringing users with the same taste and preferences together and as a result, it will reduce the grey sheep users effect because their recommended items will be based on the taste of their closest users.

In this paper, we propose a new identification approach of grey sheep users based on similarity and trust, inspired from the clustering-based work of [4]. The main idea is to compute a new weight per target user that takes into consideration these both measures and then affect it the appropriate cluster in order to enhance the quality of identification and, consequently recommendations.

This paper is organized as follows Sect. 2 is dedicated to methods solving the grey sheep problem. Section 3 details our proposed framework for identifying the grey sheep users. An illustrative example is presented in Sect. 4. Section 5 exposes the experimental study.

2 State of the Art

In this section, we will expose different identification techniques that allow solving many problems that a recommender system may face and especially identifying grey sheep users in a recommender system. They can be classified into three groups: spatial outliers detection techniques, statistical techniques and clustering techniques.

Outliers are defined as the number of observations in a set of data, which appears to be completely different with the remaining set of that data [2]. Authors in [10] proposed the Iterative z algorithm to deal with the GSU. The idea is to replace the attribute value of the outlier by the average attribute value of its neighbors and some subsequent updating computation. Another version of this algorithm, the Iterative r has been implemented using two functions a neighborhood and a comparison function to detect outliers (GSU). An other technique, the Median algorithm has also been proposed which consists in choosing the median of the attribute values of the points in order to detect outliers.

Matrix Factorisation (MF) based techniques can be classified as a linear algebra technique. In [7], authors used the MF during the identification phase in order to reduce matrix dimensions. They proposed the GSUOnly model that evaluates in what extent a grey sheep user can benefit from an other grey sheep user. They also proposed the WeightedGSU model that measures the weight of the GSU during the learning phase. In standard algorithms both GSU and normal users have the same weight. In this technique, authors considered W_{GSU} (Weight Grey Sheep User) as the weight of W_{GSU} and W_{Normal} as the weight of normal users. And finally, they proposed the SingleGSU model that forms one model for each GSU. It is designed to evaluate to what extent a model dedicated to a specific GSU improves the accuracy of the recommendations provided to this user.

The detection of GSU can also be considered as a clustering problem. In [4], authors propose the KMeansPlus which chooses randomly users to form initial centroids and then assign the rest of users depending on the distance between them. KMeansPlusPower uses the power user measure instead of distance, while KMeansPlusProbPower uses both distance and power user measures.

Recent clustering based identification approaches give satisfactory results [12]. They help a lot in giving a good idea about users' types and tastes. However, all approaches are based upon similarities between users. They only rely on users ratings, the number of items they rated and the number of neighbors, etc. These measures are no longer sufficient because any user can give a fake rating to an item which makes the users' classification bad and as a consequence decreases the accuracy of a recommender system.

3 A Novel Approach in Identifying Grey Sheep Users

With the appearance of social networks during the last few years, social trust based recommender systems have seen the light [11]. These latters take into consideration the rating provided by users that have a direct or an indirect trust relation with the active user which makes the system's learning much better and that by giving him more idea about users' relations and consequently a better vision about items they may like in the future. In general, trust is a measure of confidence that an entity or entities will behave in an expected manner. It is widely accepted as a major component of human social relationships [16].

Inspired from this idea, we propose a new GSU identification approach based on [4]. The main idea is to detect the GSU cluster based on similarity and trust measures. To this end, a new weight is computed per target user to affect it to the appropriate cluster. The proposed framework is composed of two phases: Identification phase and recommendation phase as outlined by Fig. 1. The new GSU identification approach based on [4], is denoted by $TKMeans++$.

Figure 1 shows the new proposed framework which is composed of two phases Identification phase and Recommendation phase.

3.1 Identification Phase

The identification phase aims to construct $k+1$ clusters of users. The k clusters are dedicated for white sheep users considered as usual users. The additional cluster is for the GSU. For each target user u_t, we propose to compute a weight per cluster c_i. This weight considers three kinds of measures, namely the similarity measure, the power user value and the trust measure.

The power user value represents the users who are considered as leaders because they rated a huge number of items in a recommender system. It is calculated as follows [4]:

$$P(u_i, u_j) = \frac{I_{u_i}}{I_{u_j}} \tag{1}$$

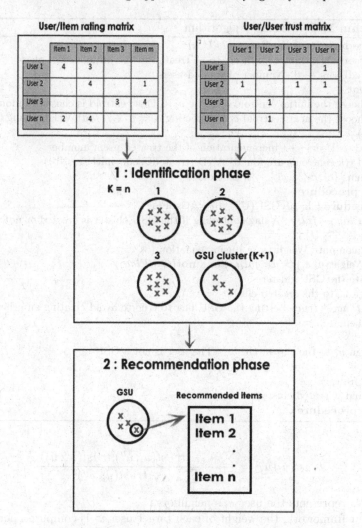

Fig. 1. The new GSU identification framework

Where $P(u_i, u_j)$ is the power user value for user u_i and u_j, I_{u_i} and I_{u_j} represent the number of items rated by user u_i and u_j, respectively.

Regarding the trust value, we choose to use the tidal trust algorithm as it is one of the most common approaches [5]. It computes implicit trust between two users who are not directly connected, who are not friends which will give a better understanding of the overall relation between users. This approach runs a depth first search to find the most direct truthful users. So, to measure the trust value among user u_i and user u_j, we need to form a collection from the trust values among the direct neighbors of u_i and u_j alloyed by the direct trust values of u and this latter directly connected neighbors. Formally, the inferred trust value is calculated as follows [5]:

Algorithm 1. TKMeans++ algorithm

1: **procedure** $InitialiseCentroids(U, k)$
2: initUsersNb – Number of users to be treated in the initialisation phase
3: treatedUsers $\leftarrow$ 0 –Number of treated users
4: **repeat**
5:      Choose the initial centroid c_1 to be u_p. –This centroid is chosen randomly
6:      Choose the next centroid c_i by selecting $c_i = u' \in U$ with weight using equation (3)
7:      treatedUsers++ –Incrementation of the treated users' number
8: **until** ((k centroids are found) AND (treatedUsers=initUsersNb))
9: **return** $\{c_1, c_2, ..., c_k\} \rightarrow$ k centroids.
10: **end procedure**
11: **procedure** ClassifyGSU(U,k,sim_{thr},GsuFlag)
12: $GsuFlag \leftarrow false$ –A flag controling if the GSU cluster is created or not
     for all($u_i \in U$)
13: $\forall\ c_i$, compute Weight(u_i,c_i) using equation (3)
14: **if** (Weight(u_i,c_i) > $weight_{thr}$) and **not**($GsuFlag$))
15: Create the GSU cluster
16: assign u_i to the created cluster
17: $GsuFlag \leftarrow true$ –Setting the GSU flag to true to avoid creating an other cluster
     for them
18: **else**
19: Assign u_i to the closest cluster –The user is not a GSU
20: **end if**
     end for
21: **return** Created clusters
22: **end procedure**

$$trust(u_i, u_j) = \frac{\sum_{u'' \in N_u} trust(u_i, u'').trust(u'', u_j)}{\sum_{u'' \in N_u} trust(u_i, u'')} \qquad (2)$$

where N_u represents the user u_i's neighbors.

In our framework, the weight of each target user u_t is computed per cluster c_i. It consists in calculating the average of similarity, power values and trust values of users pertaining to the same cluster. Formally,

$$Weight(u_t, c_i) = \frac{\frac{\sum_{u_i \in c_i} sim(u_t, u_i)}{size_{c_i}} + \frac{\sum_{u_i \in c_i} P(u_t, u_i)}{size_{c_i}} + \frac{\sum_{u_i \in c_i} trust(u_t, u_i)}{size_{c_i}}}{3} \qquad (3)$$

Where $sim(u_t, u_i)$ is the similarity between users, $P(u_t, u_i)$ is the power user value expressed by Eq. (1), $trust(u_t, u_i)$ is the trust value between u_t and u_i expressed by Eq. (2), respectively and $size_{c_i}$ is the number of users in the current cluster c_i.

The obtained weight is then compared to a defined threshold $weight_{thr}$ and consequently, the target user is affected to the appropriate cluster. If it is less that $weight_{thr}$, the user is considered as a white sheep user and assigned to the closest cluster. In the opposite case, we deal with a grey sheep user. Algorithm 1 outlines the TKMeans++ identification process. The complexity of our proposed method is $O(n^2)$.

As in the classical KMeans [8], we choose at the beginning k random centroids. The first research question in a clustering algorithm is: "Does the centroids initialization have an impact on the accuracy of the clustering?". In [1], the researchers proposed the KMeans++ algorithm that uses a probabilistic approach in order to select the initial centroids and proved that it yields to a much finer clustering. Inspired from this idea, we add trust relations to the same process of [4]. The idea is to randomly affect a user to each cluster. Then, we take a set of users who will be initially affected to clusters according to their weight values expressed by Eq. (3).

3.2 Recommendation Phase

After identifying the different users of our system, the recommendation phase takes place by exploiting the GSU cluster produced in the identification phase in order to help the recommender system predicting the users' future interests.

Our purpose is to calculate the preference of the user u_t based on the users that belong to the GSU cluster. By doing this, the recommendation is restricted on the items rated by users belonging to the same cluster as the active user.

In fact, instead of computing user's preferences, taking into consideration only the movies that are well evaluated in the active user's corresponding cluster will make the recommendation phase much more pertinent and efficient in terms of execution time because the movies that are evaluated negatively by the active user's cluster and that are not watched by this latter will not be taken into consideration.

To compute the predicted active user's rating on each item, we take the average rating given by users in the GSU cluster. The rating prediction of item i for target user u_t is calculated as follows:

$$P_{u_t,i} = \overline{r_{u_t}} + \frac{\sum_{u_j \in c_{GSU}} (r_{u_j,i} - \overline{r_{u_j}}) * sim(u_t, u_j)}{\sum_{u_j \in c_{GSU}} |sim(u_t, u_j)|}, \forall u_j \in c_{GSU} \qquad (4)$$

where u_j a grey sheep user, c_{GSU} the GSU cluster, $r_{(u_j,i)}$ is the rating given by u_j on i and $\overline{r_{u_i}}$ is their average ratings for user u_i.

The top n recommended items' list for the active user includes the n candidate items that have the highest predicted preferences.

4   Illustrative Example

Let us consider the users/items matrix described by Table 1. We have 5 users and 5 movies, each user can give a rating to a movie from 1 to 5. The non-assigned ratings mean that the user u_i did not watch the movie M_i or did not rate it.

Table 1. Example of a rating matrix

$User/Movie$	$M1$	$M2$	$M3$	$M4$	$M5$
u_1	5	4	–	3	3
u_2	4	–	3	5	–
u_3	3	3	3	–	5
u_4	1	1	–	–	2
u_5	1	–	1	2	1

Table 2 represents a user/user trust matrix. We have 5 users who can trust or not trust an other users. The trust value equal to 1 means that u_i trusts u_j, 0 otherwise.

Initially, we fix the number of users who are going to be treated during the initialisation phase to 3 (users who will construct the initial centroids), the number of centroids k is set to 2 and the $Weight_{thr}$ to 0.6.

At first, we assume that u_1 pertains to cluster c_1 and then calculate the weight for u_2. We obtain Weight$(u_2, c_1) = 0.43 < Weight(u_2, c_2) = 0.55$, so u_2 will be assigned to c_2. The same process is done for u_3, which is affected to c_1.

To calculate the Power user measure:

Table 2. Example of trust matrix

$User/Movie$	u_1	u_2	u_3	u_4	u_5
u_1	–	1	–	–	–
u_2	1	–	1	–	–
u_3	–	1	–	–	–
u_4	–	–	–	–	1
u_5	–	–	–	1	–

$$P(u_1, u_3) = \frac{\text{Number of movies rated by } u_1}{\text{Number of movies rated by } u_3} = \frac{1}{1} = 1 \tag{5}$$

After the initialisation phase, we start the rest of users' classifcation, namely u_4 and u_5. We calculate the weight value for u_4 and the cluster c_1 and find out that it exceeds the $Weight_{thr}$.

$$Weight(u_4, c_1) = \frac{\frac{\sum_{u_i \in c_i} sim(u_4, u_i)}{size_{c_1}} + \frac{\sum_{u_i \in c_i} P(u_4, u_i)}{size_{c_1}} + \frac{\sum_{u_i \in c_i} TidalTrust(u_4, u_i)}{size_{c_1}}}{3}$$

$$Weight(u_4, c_1) = \frac{\frac{(-0.56 + -0.77)}{2} + \frac{(2+1)}{2} + \frac{2}{2}}{3} = 0.61$$

As a result, the c_{GSU} is created and $GsuFlag$ is set to true and u_4 is assigned to this created cluster because he is far away from both created clusters so he is identified as a GSU and is isolated in a unique cluster in order to help identifying users having the same behaviors than him and grouping them in a same place to learn their tastes more efficiently.

Finally, we calculate the weight value for u_5 and assign it to c_{GSU} because its weight comparing to this cluster exceeds its weight comparing to the other ones (Weight($u_5, c_1 = \{u_1\}$) = 0.19, (Weight($u_5, c_2 = \{u_2\}$) = 0.12, (Weight(u_5, $c_{GSU} = \{u_4\}$) = 1.22). Now, we have 3 clusters, $c_1 = \{u_1, u_3\}$, $c_2 = \{u_2\}$ and $c_{GSU} = \{u_4, u_5\}$. It is clear from the Table 1 that u_4 and u_5 are GSU. Their ratings are not in agreement with the rest of the users present in the table (they almost disliked all the items).

Let us compute the rating of M_3 and M_4 for GSU u_4. We find that $P_{u_4, M_3} = 0.9$ and $P_{u_4, M_4} = 1.6$. In this case, the system will recommend the top-2 movies M_4 first and then the movie M_3.

5   Experimental Study

To evaluate TKMeans++, we will use the Epinions[1] dataset which is a website where people can review products. It contains 131,228 users, 317,755 items and 1,127,673 reviews. 113,629 users have slightly one rating, 47,522 users have slightly one trust relation and 4,287 users have neither reviews nor trust relation. We randomly divise our dataset into two sets namely: training set (80%) and testing set (20%). We execute our algorithm over 1000 users (chosen randomly) from the Epinions dataset during the centroids' initialisation phase.

In order to evaluate the quality of the provided recommendations, we choose the classification accuracy metric coverage expressed by Eq. 6 and the predictive accuracy metric MAE expressed by Eq. 7. We focus on comparing our proposed method TKMeans++ with the algorithms KMeans, KMeansPlus, KMeansPlus-Power and KMeansPlusProbPower proposed in [4].

Coverage: The coverage of a recommender system is a measure of the domain of items in the system over which the system can form predictions or make recommendations. Systems with lower coverage may be less valuable to users, since they will be limited in the decisions they are able to help with.

$$Coverage = \frac{\sum_{u \in U^{test}} \sum_{i \in i_u^{test}} 1R_{>0}(r_{i,u})}{N} \qquad (6)$$

Where U^{test} represents the testing set users, i_u^{test} represents the items evaluated by user u in the testing set, $R > 0$ represents the set of real numbers exceeding zero, N is the number of rating records in the testing set and $1R_{>0}(r_{i,u})$ is an index, which is expressed as follows:

$$1R_{>0}(r_{i,u}) = \begin{cases} 1 \text{ if } r_{i,u} \in R > 0, \\ 0 \text{ otherwise.} \end{cases}$$

---

[1] https://projet.liris.cnrs.fr/red/.

Mean Average Error (MAE):

MAE calculates the mean absolute deviation in a predictions' collection. It is the mean over the test set of the overall diversity between the prediction and the real evaluations where all users' inequalities have equal weight.

$$MAE = \frac{\sum_{i=1}^{N} |P_{u,i} - R_{u,i}|}{N} \tag{7}$$

Where $P_{u,i}$ denotes the predicted rating, $R_{u,i}$ denotes the real ratings and N represents the total number of users/items that are going to be predicted.

Before presenting our experimental results, we should fix $Weight_{thr}$, the threshold that determines and controls the number of grey sheep users. It represents the key of our algorithm as it is useful for the assignment of a target user to the GSU cluster. To this end, we changed the value of $Weight_{thr}$ and measured the corresponding MAE of the users not identified as grey sheep users. The value that gives the minimum MAE is termed as the optimal threshold value of $Weight_{thr}$. After running several experiments, we fix it to 361.

For finding the optimal number of clusters, we changed the cluster size from 10 to 200 with a difference of 10, and measured the corresponding MAE, over the training set. We find that 90 is the optimal value of k.

In what follows, we present the MAE and coverage results over the different number of clusters and $Weight_{thr}$ in order to find the optimal value of both parameters.

Figure 2 shows how the MAE changes with an increase in the number of clusters. For the Epinions dataset, the MAE keeps on decreasing with an increase in the number of clusters until it converges. To keep a good balance between computation and performance requirement, we choose the optimal number of clusters to be 90.

Figure 3 shows how the coverage results are changing with an increase in the number of clusters. For the Epinions dataset, we observe that the coverage is maximum at k = 90.

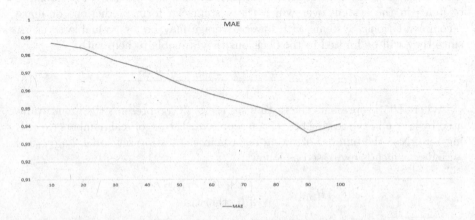

Fig. 2. Finding the optimal number of clusters in terms of MAE.

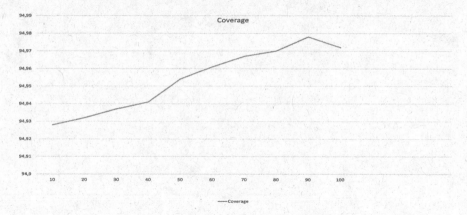

Fig. 3. Finding the optimal number of clusters in terms of coverage.

We observe from Fig. 4 that the MAE decreases is related to the increase of the number of clusters. For the Epinions dataset, the Fig. 4 shows that the minimum MAE value is obtained with a $Weight_{thr}$ value equal to 361.

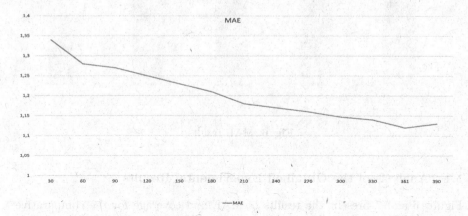

Fig. 4. Finding the optimal Weight threshold in terms of MAE.

Figure 5 shows the different coverage results within the increase in terms of clusters' numbers. For the Epinions dataset, we observe that the coverage keeps increasing proportionnaly with the value of the $Weight_{thr}$ and reaches its maximal value when the $Weight_{thr}$ is equal to 300.

We choose a number of clusters equal to 361 to run our experiments. This value is chosen in order to guarantee a minimum MAE value which means that the recommendations given by the TKmeans++ are more accurate. Also, 361 is not far from the optimal number of clusters found for the coverage measure.

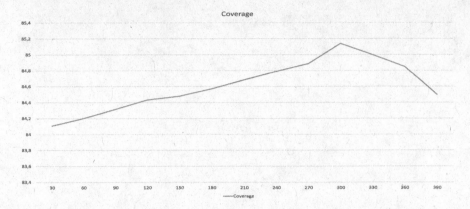

Fig. 5. Finding the optimal Weight threshold in terms of coverage.

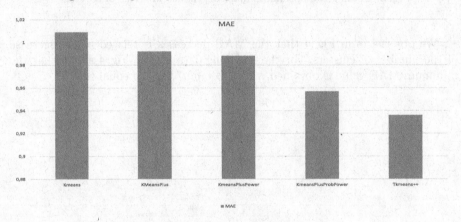

Fig. 6. MAE results.

5.1   Analysis of the Obtained Experimental Results

Figures 6 and 7 present the results of MAE and coverage for the comparative methods. From Fig. 6, we can pinpoint that TKmeans++ algorithm exceeds the classical KMeans by 7.235%, KMeansPlus by 5.645%, KMeansPlusPower by 5.263% and KMeansPlusProbPower by 2.194% in terms of MAE. This means that TKmeans++ makes less error in recommending items to the grey sheep users and it is due to the weight expressed by Eq. 3. In other terms, taking into consideration trust relations between users has a positive impact on recommendation and this by generating less error.

Regarding the coverage presented in Fig. 7, it is obvious that TKmeans++ and KMeansPlusProbPower methods have approximately the same value. This means that both methods can be valuable for GSU users by providing a percentage of 95%. For the other methods, TKmeans++ exceeds KMeans by 1.502%, KMeansPlus by 0.396% and KMeansPlusPower by 0.494%.

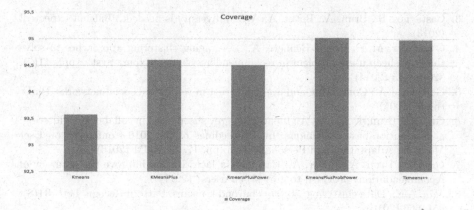

Fig. 7. Coverage results.

We can finally conclude that inferring trust in collaborative filtering provides better results, notably, when we infer trust values for identifying GSU because it will surely make recommendations more pertinent and refine them.

6   Conclusion

This paper deals with the GSU problem which concerns users whom opinions do not match with any group of users and, therefore, are unable to get benefit from collaborative filtering. In other words, the grey sheep users are a group of users who may neither agree or disagree with the vast majority of the system's users. They may represent many difficulties in producing accurate recommendations which degrades the quality of the recommender system output.

Throughout this paper, we have shown that taking only into account the distance between users in classifying them is not good enough, we have shown that trust relations between users help a lot the algorithm in the understanding of users' behaviors and tastes and in learning the good way in grouping them. As a result of this, adding trust factors will help in bringing users more close to each other, hence, will improve the quality of recommendations given to GSU users.

As a future work, we will explore other model-based GSU approaches like matrix factorization and study the merging of different models and its impact for the GSU problem.

References

1. Arthur, D., Vassilvitskii, S..: k-means++: The advantages of careful seeding. In: Proceedings of the Eighteenth Annual ACM-SIAM Symposium on Discrete Algorithms, pp. 1027–1035. Society for Industrial and Applied Mathematics (2007)
2. Barnett, V., Lewis, T.: Outliers in Statistical Data. Wiley, New York (1994)

3. Castagnos, S., Brun, A., Boyer, A.: When diversity is needed... but not expected! (2013)
4. Ghazanfar, M.A., Prugel-Bennett, A.: Leveraging clustering approaches to solve the gray-sheep users problem in recommender systems. Expert Syst. Appl. **41**(7), 3261–3275 (2014)
5. Golbeck, J.A.: Computing and applying trust in web-based social networks. Ph.D. thesis (2005)
6. Gras, B., Brun, A., Boyer, A.: Identifying grey sheep users in collaborative filtering: a distribution-based technique. In: Proceedings of the 2016 Conference on User Modeling Adaptation and Personalization, pp. 17–26. ACM (2016)
7. Gras, B., Brun, A., Boyer, A.: Can matrix factorization improve the accuracy of recommendations provided to grey sheep users? (2017)
8. Jain, A.K.: Data clustering: 50 years beyond k-means. Pattern Recogn. Lett. **31**(8), 651–666 (2010)
9. Lopez-Nores, M., Blanco-Fernandez, Y., Pazos-Arias, J.J., Gil-Solla, A.: Property-based collaborative filtering for health-aware recommender systems. Expert Syst. Appl. **39**(8), 7451–7457 (2012)
10. Lu, C.T., Chen, D., Kou, Y.: Algorithms for spatial outlier detection. In: Third IEEE International Conference on Data Mining, pp. 597–600. IEEE (2003)
11. Massa, P., Avesani, P.: Trust-aware recommender systems. In: Proceedings of the 2007 ACM Conference on Recommender Systems, pp. 17–24. ACM (2007)
12. Merialdo, A.K.B.: Clustering for collaborative filtering applications. Intell. Image Process. Data Anal. Inf. Retrieva **13**, 199 (1999)
13. Resnick, P., Iacovou, N., Suchak, M., Bergstrom, P., Riedl, J.: GroupLens: an open architecture for collaborative filtering of netnews. In: Proceedings of the 1994 ACM Conference on Computer Supported Cooperative Work, pp. 175–186. ACM (1994)
14. Zheng, Y., Agnani, M., Singh, M.: Identification of grey sheep users by histogram intersection in recommender systems. In: Cong, G., Peng, W.-C., Zhang, W.E., Li, C., Sun, A. (eds.) ADMA 2017. LNCS (LNAI), vol. 10604, pp. 148–161. Springer, Cham (2017). https://doi.org/10.1007/978-3-319-69179-4_11
15. Zheng, Y., Agnani, M., Singh, M.: Identifying grey sheep users by the distribution of user similarities in collaborative filtering. In: Proceedings of the 6th Annual Conference on Research in Information Technology, pp. 1–6. ACM (2017)
16. Sherchan, W., Nepal, S., Paris, C.: A Survey of Trust in Social Networks. IBM Research- Australia, CSIRO ICT Centre

Digital Marketing

Opportunity for Video-on-Demand Services – Collecting Consumer's Neurophysiology Data for Recommendation Systems Improvement

Kristian Dokic[1,2](✉) and Tomislava Lauc[1,2]

[1] Polytechnic in Pozega, Vukovarska 17, 34000 Pozega, Croatia
kdjokic@vup.hr, tlauc@ffzg.hr
[2] Faculty of Humanities and Social Sciences, Organization Ivana Lucica 3,
10000 Zagreb, Croatia

Abstract. In the last few decades, the way that consumers watch video content has changed. Video-on-demand services usage has been raised, and this enables some new opportunity to improve video content recommendation systems of those services. New video-on-demand services usually use the Internet as a broadcast infrastructure so the Internet can be used for feedback sending. Feedback can be divided into two groups: based on conscious choices generated by the consumer or based on consumer's neurophysiological data. In this paper, the second option is analyzed, and the focus is on gender differences. Participants have watched four movie trailers, and different neurophysiology data have been recorded while they have been watching the trailers. During that time, they have been rating trailers. Heart rate and galvanic skin response have been extracted and analyzed in different ways. A weak correlation between trailers scores and standard deviation of heart rate was detected. Still, on the other hand, a statistically significant difference in the numbers of detected skin conductance responses between the genders was measured from the sample. This knowledge could be implemented in the rating systems for further improvement. Also, the use of consumer's neurophysiological data in the video-on-demand services rating systems should be further investigated.

Keywords: Video-on-demand services · EDA · GSR · Heart rate · Neurophysiology data · Recommendation systems

1 Introduction

Recommending systems help video-on-demand services consumers to identify movies or video content that are more attractive to them. These systems can be divided into two major types: personalized and non-personalized recommendation systems [1]. Non-personalized proposes movie or video without consumer profiles analysis. This paper deals with one idea that can improve personalized recommendation systems. Their recommendations are based on different elements like individual preferences or choices, demographic information, etc. Recommending systems provide benefits to consumers and services. Consumers can be sometimes overwhelmed by too much content, and RSs

© Springer Nature Switzerland AG 2020
M. A. Bach Tobji et al. (Eds.): ICDEc 2020, LNBIP 395, pp. 91–104, 2020.
https://doi.org/10.1007/978-3-030-64642-4_8

help them to identify movie or video content of their interest efficiently. On the other hand, video-on-demand services can influence consumer attitude and decisions.

In this paper, a short review of the literature about recommendation systems will be provided, but the main idea of the paper is to use neurophysiology data to improve these systems. In the second section literature about electrodermal activity and heart rate will be provided. In the third section, the research will be described. In the first part, sample and methodology, and the second part, data processing and results. Discussion is in the fourth section, and finally, the conclusion is in the fifth section.

The focus of the research is on heart rate and electrodermal activity data analysis at the time of movie trailers watching to identify gender differences. It has been chosen to prove that recorded neurophysiological responses contain valuable information that could enhance a recommendation systems model. This is a rather narrow area where no research is available, but on the other hand, the analysis of neurophysiology data is quite well documented.

2 State of the Art

2.1 Recommendation Systems Based on Conscious Choices

There are different types of recommendation systems based on knowledge source. Burke divide them into the four classes:

a) Collaborative,
b) Content-based,
c) Demographic,
d) Knowledge-based [2, 3].

Soares and Viana presented their web-based recommendation system that is based on collaborative and content-based filtering. Their algorithm enables interoperability between traditional broadcast and video-on-demand services [4].

Pelaja et al. have proposed an innovative algorithm and the framework for movie ratings based on sentiment analysis of user comments. They included sentiment knowledge from forum discussion about movies, and they improved recommendations accuracy [5].

Gupta and Moharir suggested that recommendation systems can be used as a base for prefetching content, especially if Video on Demand service has servers around the world. The main goal of their paper is to optimize network resources by modelling time-correlation in consumer requests [6, 7]. Verhoeyen et al. have analyzed the same topic, and they conclude that the clever usage of user's content preferences can exceed the performance of caching algorithms that are usually based on popularity [8].

Guntuku et al. suggested that the user interface has to be personalized to get more quality results from Video-on-Demand recommendation system. They proposed three personalized user interfaces for different consumer behaviour: channel surfer, recorder and cast fun [9].

Mo et al. proposed a cloud-assisted recommendation system that stores recommendation rules in the cloud. Authors used comment information for rules generating, and they achieve better results compared to standard recommendation systems which store recommendation lists [10]. Tsunoda and Hoshino concluded that more metadata about items is needed to improve the quality of recommendation systems [11].

All of them are based on conscious choices made by individuals. On the other hand, recommendation systems can also get some psychophysiological data from a person and maybe improve the quality of recommendation. There are lots of psychophysiological data that can be measured, but in this paper, the focus will be on electrodermal activity and heart rate.

2.2 Electrodermal Activity

The idea of skin conductance measuring is from the 19th century. Still, first serious researches that deal with electrodermal activity are from the first half of the twentieth century, so there is a lot of experience on that field. The electrodermal activity has a complex mechanism with lots of factors interacting, and that was concluded by Coombs before almost eighty years [12, 13].

There is a little confusion about term *Electrodermal activity* because there are a lot of synonyms because of a long history of research on that field. Some of them are galvanic skin response (GSR), electrodermal response (EDR), skin conductance response (SCR) or only skin conductance (SC). In this paper and generally, standardized term *Electrodermal activity* is mostly used [12].

Skin conductance varies because of skin sweating that is controlled by a sympathetic nervous system. Sweat gland activity increases if the sympathetic nervous system is highly aroused. Skin conductance rises as a result of sweating [14].

Healey has measured drivers skin conductance and concluded that a reliable system for driver stress recognition could be made with only two sensors, galvanic skin response and electrocardiogram sensors [15]. There are lots of other authors that used skin conductance for stress detecting [16–18].

Chanel et al. measured the different peripheral physiological activity of game players, and they have noticed decreasing in galvanic skin response when the game difficulty has increased [19].

Money and Agius measured electrodermal activity in response to different video content. They used action, science fiction, drama, horror and comedy. They also presented an analysis framework for psychophysiological data. They concluded that horror, action and science fiction movies significant elicit electrodermal response unlike comedy and drama, that evokes a much lower level of electrodermal response [20].

Soleymani et al. presented the research with eight participants that watched 64 different scenes from eight movies. Their psychophysiological responses (EDA, EMG, blood pressure, skin temperature and respiration pattern) were recorded while they watched these scenes. After watching each scene, participants filled the questionnaire about valence and arousal about the scene. Authors calculated the positive correlation between valence and arousal rankings from surveys and data from psychophysiological responses [21].

Ali et al. proposed globally generalized emotion recognition system based on Cellular Neural Networks with accuracy between 80% and 89%. They analyzed ECG, EDA and skin temperature to recognize four different emotions [22].

2.3   Heart Rate

Heart rate is an essential indicator of the consumer's emotional state. There are different ways to measure heart rate, but generally, it is measured using the average time between peaks of measured waves. Heart rate is accelerated high after some emotional stress, loud noise or doing some exercise [23]. Some authors suggest that low arousal is associated with no heart rate acceleration or deceleration. Still, high arousal is associated with heart rate acceleration and deceleration and generally high heart rate [24].

Lisetti and Nasoz used heart rate and other physiological signals from the autonomic nervous system to discover different emotions (anger, surprise, amusement, frustration, fear and sadness). They used three different supervised machine learning algorithms, and they reach very high accuracy. For anger, they rich 91,7% accuracy with MBP algorithm [25].

Santos Sierra et al. proposed a stress detection system based on galvanic skin response and heart rate data that can reach more than 99% stress detection accuracy in only ten seconds [26].

Some authors used ECG data in their research because ECG includes more information than a simple heart rate. Shalini and Vanitha proposed a model for three different emotions detection based on ECG data. They used fast Fourier transformations to remove noise from the ECG signal and then sadness, fear and joy can be efficiently identified [27].

Tan et al. developed a model for a personal affective trait, and they did not use heart rate data because it was strongly correlated with arousal in all cases. Because of the personalization, they had to exclude heart rate data [28].

3   Research

3.1   Sample and Methodology

In this research, students have watched movie trailers, and differences between the female and male group in electrodermal activity and heart rate have been analyzed. Four trailers have been selected for research. Two trailers have been recorded from action movies genre ("No time to die" and "Underwater") and another two from romance movies genre ("I still believe" and "Little women"). They have been selected from the Internet Movie Database, and the trailers from the movies that were not available in the theatres have been chosen intentionally. These movies will be available in the cinemas around the world during 2020. Durations of trailers are:

a) "I still believe" - 152 s
b) "No time to die" - 165 s
c) "Little women" - 163 s
d) "Underwater" - 138 s

Some authors suggested that this is sufficient time for recording neurophysiological responses [29]. All trailers have been linked in one video clip that lasts 618 s. This video is available on YouTube service (https://www.youtube.com/watch?v=qB2JRZVz3AE).

Twenty-one polytechnic students (11 females and 10 male) aged between 20 and 33 years old were included in the research. They were watching video clip with linked four trailers and they had headphones and Empatica E4 wrist band for psychophysiological data recording. After every trailer student rated the trailer in the questioner with a score between one and ten. Beside the digit 1 on the questioner there is the text "I do not want to watch!" and beside the digit 10 is the text "I want to watch!".

Fig. 1. Empatica E4

Empatica E4 wrist band is used in the research. It can be seen on Fig. 1, and it is equipped with sensors for skin temperature, blood volume pulse, motion and fluctuation changes in electrical properties of the skin (Electrodermal Activity). These sensors collect data from the sympathetic branch of the nervous system, including the blood volume pulse (BVP), galvanic skin response (GSR), heart rate (HR), acceleration and skin temperature (ST). There are lots of papers available that used this device [30–33].

Empatica E4 wrist band have to be connected with a smartphone with Bluetooth connection while data is recording. Application for data recording can be downloaded from Google Play service. On the Fig. 2 acquisition in progress can be seen.

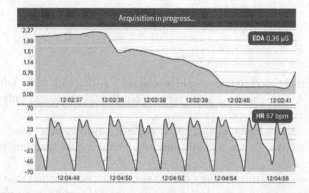

Fig. 2. Empatica E4 Android application (acquisition in progress)

3.2 Data Processing and Results

After data recording, there were data from the questionnaire and data from Empatica cloud available. Data from the questionnaire are in Table 1.

Table 1. Data from the questionnaire

Student	"I still believe"	"No time to die"	"Little women"	"Underwater"	M = 0/F = 1
1	7	8	5	6	1
2	10	8	8	10	1
3	9	8	5	10	1
4	10	7	5	10	1
5	10	7	5	10	0
6	10	1	1	1	1
7	3	5	5	8	1
8	9	6	5	5	0
9	5	10	5	8	0
10	3	9	3	10	0
11	10	1	1	1	0
12	8	4	7	4	0
13	1	7	2	7	0
14	1	10	1	5	0
15	10	9	10	10	1
16	10	9	9	5	1
17	10	6	8	6	1
18	10	8	9	8	1
19	10	8	3	3	1
20	4	10	1	1	0
21	8	8	6	7	0

From the Empatica cloud, we could download six files with some kind of data for every student. All files were in CSV format, but for this research, only HR.csv and EDA.csv are used.

Files with heart rate data, named HR.csv, contain average heart rate frequency in a minute, and it is based on ten seconds intervals. Frequency of sampling is one sample per second.

Files with electrodermal activity data, named EDA.csv, contain skin conductance values sampled four times per second. For ten minutes' recording, there were about two thousands and five hundred samples per student.

In Table 2, there are main descriptive statistics of all student's heart rate data.

Table 2. Descriptive statistics of heart rate

Mean	80,33
Median	81,25
Mode	78,07
Standard deviation	10,74
Minimum	53,13

After different data analysis, a weak negative correlation between trailers scores and standard deviation of heart rate has been found in the female group data (R = -0,30; p = 0,050). On the other hand, a weak positive correlation between trailers/movie scores and standard deviation of heart rate has been found in the male group data (R = 0,32; p = 0,047). It can be seen in Figs. 3 and 4.

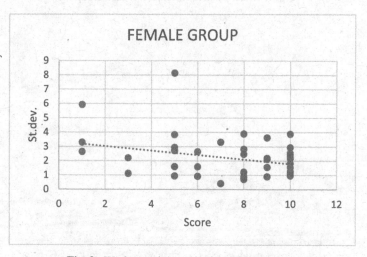

Fig. 3. Weak negative correlation – female group

Raw data from EDA.csv file could not be statistically analysed without previous preprocessing. Raw data from EDA.csv can be seen on the chart on the Fig. 5, so it is evident that there is a big difference between recorded values.

Matlab based software/plugin called Ledalab has been used, and it is usually used for electrodermal activity data preprocessing in researches. It can import various file formats and provide many preprocessing functions. In Fig. 6, Ledalab output windows after data preprocessing can be seen. In the first chart, raw data can be seen, but in the third one, there is raw data without a tonic component. There are a lot of similar researches that have been used this software/plugin for preprocessing [34–37].

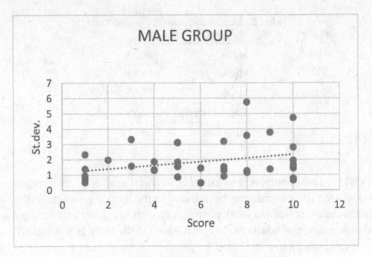

Fig. 4. Weak positive correlation – male group

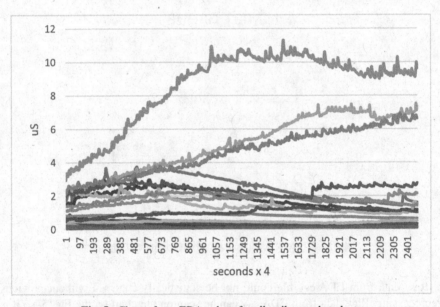

Fig. 5. Chart of raw EDA values for all trailers and students

In this research, Ledalab has been used for decomposition of skin conductance into tonic and phasic component. Ledalab provides discrete decomposition analysis, and it has been used. Benedek proposes this method and it explores deviations of response shape and computes a model of all components [38].

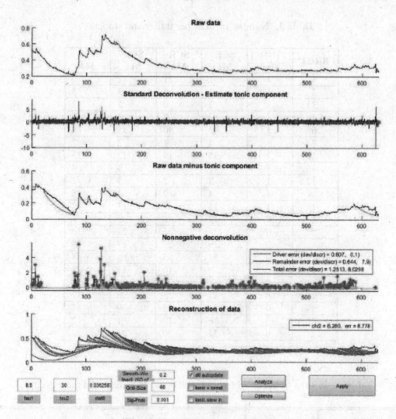

Fig. 6. Ledalab output window

Data from every student have been analysed with Edalab and after that twenty-one output file with a list of detected skin conductance response have been available. The list for every student is a simple series of numbers that represent the number of seconds when the conductance response is observed. Values for every trailer is grouped so final results can be seen in Table 3. In the first and second row, there are numbers and gender of students. From the third to sixth row, there are numbers of detected skin conductance responses for every student-trailer combination. Last ten rows are colored with a light grey because these students are male.

Difference between female and male students has been analyzed with t-test, and results are in Table 4.

Difference between female and male group can also be seen on the chart on the Fig. 7.

Table 3. Number of SCR per trailer and student

BROJ	F=1; M=0	SCR M1	SCR M2	SCR M3	SCR M4
1	1	31	21	12	9
2	1	2	4	3	3
3	1	2	0	1	2
4	1	7	7	3	5
6	1	41	40	35	28
7	1	4	1	2	2
15	1	35	19	17	15
16	1	1	3	3	4
17	1	2	2	3	2
18	1	4	2	3	5
19	1	15	7	13	12
5	0	13	8	5	5
8	0	27	26	23	19
9	0	29	25	17	13
10	0	51	55	50	48
11	0	9	16	11	21
12	0	34	23	25	21
13	0	27	23	26	12
14	0	41	45	39	31
20	0	6	3	1	2
21	0	33	30	21	22

Table 4. Results of t-test – the difference between female and male group

		Levene's Test for Equality of Variances		t-test for Equality of Means						
									95% CONF	
		F	Sig.	t	df	Sig. (2-tailed)	Mean Difference	Std. Error Difference	Lower	Upper
SCR	Equal variances assumed	1,47	,229	4,83	82,00	,000	13,58	2,81	7,99	19,18
	Equal variances not assumed			4,78	74,99	,000	13,58	2,84	7,92	19,24

Lots of different analysis have been performed with the data, but because of the low number of observations, the level of significance was not acceptable. We used Schwarz's Bayesian criterion for variable selection, but mostly they were insignificant [39]. ANCOVA has also been used with different combinations but without acceptable levels of significance.

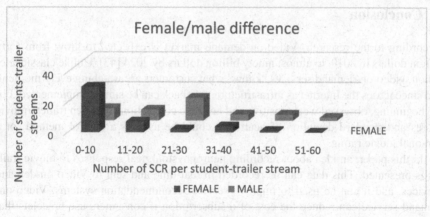

Fig. 7. Difference between female and male groups

4 Discussion

In this paper, the new opportunity for improving recommendation systems of video-on-demand services is suggested. The idea is not new, because there are lots of papers that discover different emotions by neurophysiological data analyzing. Still, as we know, no one video-on-demand service use neurophysiological data to improve recommendation system.

Most of the recommendation systems today use some individual preferences to suggest some new product or movie, but these preferences are based on conscious choices generated by the consumer. We recommend that neurophysiological data contains information about individual preferences that are still hidden to video-on-demand services and have to be discovered.

In this paper, the focus was on gender differences, and we wanted to find is there any differences in heart rate and electrodermal activity data between genders. Deng et al. suggested that women exhibited a significantly smaller decline in heart rate while watching videos that induce anger, pleasure and amusement. Still, on the other hand, Rohrman et al. suggested that heart rate increased more strongly in the female group than in the male group while watching videos that induce disgust [40, 41]. These results cannot be used to compare with our, but we also found a difference between the male and female group. We found a weak negative correlation between trailers scores and standard deviation of heart rate in the female group data (R = −0,30; p = 0,050). On the other hand, a weak positive correlation between trailers/movie scores and standard deviation of heart rate has been found in the male group data (R = 0,32; p = 0,047).

The difference in electrodermal activity between the female and male group is also statistically significant but only in skin conductance response number between genders, regardless of the trailer genre. The male group has a statistically significant higher number of skin conductance response. In literature, similar results can be found in Alexander and Wood paper, where they suggested that male group has higher levels of electrodermal response than the female group at resting levels [42].

5 Conclusion

According to the research, Video-on-demand market is expected to grow from forty billion dollars in 2019. to almost ninety billion dollars by 2024 [43]. Unlike classic television, video-on-demand services "know" what customers are watching every moment, and since it uses the Internet as infrastructure, feedback can be simply implemented. For the beginning, observed weak positive and negative correlations between trailers/movie scores and standard deviation of heart rate could be used as a control method for a personal movie rating.

In this paper, an idea about recording neurophysiological responses to movie trailers is presented. This data can be recorded in real-time and sent to Video-on-demand services, and it can be used to improve service recommendation systems. Video-on-demand service can reduce the cost of a subscription if consumers use bracelets that record neurophysiological responses. To prove our idea, we used the data obtained from one such device to show that it contained valuable information that could enhance the recommendation systems model. Our findings regarding correlations between trailers scores and standard deviation of heart rate, as well as the difference in electrodermal activity between the female and male group, can be used to improve a recommendation system model.

References

1. Ricci, F., Rokach, L., Shapira, B.: Introduction to recommender systems handbook. In: Ricci, F., Rokach, L., Shapira, B., Kantor, P. (eds.) Recommender Systems Handbook, pp. 1–35. Springer, Boston (2011). https://doi.org/10.1007/978-0-387-85820-3_1
2. Burke, R.: Hybrid recommender systems: survey and experiments. User Model. User-Adap. Inter. **12**, 331–370 (2002)
3. Burke, R.: Hybrid web recommender systems. In: Brusilovsky, P., Kobsa, A., Nejdl, W. (eds.) The Adaptive Web, pp. 377–408. Springer, Heidelberg (2007). https://doi.org/10.1007/978-3-540-72079-9_12
4. Soares, M., Viana, P.: TV recommendation and personalization systems: integrating broadcast and video on-demand services. Adv. Electr. Comput. Eng. **14**, 115–120 (2014)
5. Peleja, F., Dias, P., Martins, F., Magalhães, J.: A recommender system for the TV on the web: integrating unrated reviews and movie ratings. Multimedia Syst. **19**(6), 543–558 (2013). https://doi.org/10.1007/s00530-013-0310-8
6. Gupta, S., Moharir, S.: Modeling request patterns in VoD services with recommendation systems. In: International Conference on Communication Systems and Networks (2017)
7. Gupta, S., Moharir, S.: Request patterns and caching for VoD services with recommendation systems. In: 2017 9th International Conference on Communication Systems and Networks (COMSNETS) (2017)
8. Verhoeyen, M., De Vriendt, J., De Vleeschauwer, D.: Optimizing for video storage networking with recommender systems. Bell Labs Tech. J. **16**, 97–113 (2012)
9. Guntuku, S.C., Roy, S., Lin, W., Ng, K., Keong, N.W., Jakhetiya, V.: Personalizing user interfaces for improving quality of experience in VoD recommender systems. In: 2016 Eighth International Conference on Quality of Multimedia Experience (QoMEX) (2016)
10. Mo, Y., Chen, J., Xie, X., Luo, C., Yang, L.T.: Cloud-based mobile multimedia recommendation system with user behavior information. IEEE Syst. J. **8**, 184–193 (2014)

11. Tsunoda, T., Hoshino, M.: Automatic metadata expansion and indirect collaborative filtering for TV program recommendation system. Multimedia Tools Appl. **36**, 37–54 (2008)
12. Boucsein, W.: Electrodermal Activity. Springer, Heidelberg (2012). https://doi.org/10.1007/978-1-4614-1126-0
13. Coombs, C.H.: Mathematical biophysics of the galvanic skin response. Bull. Math. Biophys. **3**, 97–103 (1941)
14. Martini, F., et al.: Fundamentals of Anatomy & Physiology, vol. 7. Pearson Benjamin Cummings, San Francisco (2006)
15. Healey, J.A.: Wearable and automotive systems for affect recognition from physiology (2000)
16. Huysmans, D., et al.: Unsupervised learning for mental stress detection-exploration of self-organizing maps. In: Proceedings of Biosignals 2018, vol. 4, pp. 26–35 (2018)
17. Ollander, S., Godin, C., Campagne, A., Charbonnier, S.: A comparison of wearable and stationary sensors for stress detection. In: 2016 IEEE International Conference on Systems, Man, and Cybernetics (SMC) (2016)
18. Smets, E., et al.: Comparison of machine learning techniques for psychophysiological stress detection. In: International Symposium on Pervasive Computing Paradigms for Mental Health (2015)
19. Chanel, G., Rebetez, C., Bétrancourt, M., Pun, T.: Boredom, engagement and anxiety as indicators for adaptation to difficulty in games. In: Proceedings of the 12th International Conference on Entertainment and Media in the Ubiquitous Era (2008)
20. Money, A.G., Agius, H.: Analysing user physiological responses for affective video summarisation. Displays **30**, 59–70 (2009)
21. Soleymani, M., Chanel, G., Kierkels, J.J.M., Pun, T.: Affective ranking of movie scenes using physiological signals and content analysis. In: Proceedings of the 2nd ACM Workshop on Multimedia Semantics (2008)
22. Ali, M., Machot, F.A., Mosa, A.H., Kyamakya, K.: CNN based subject-independent driver emotion recognition system involving physiological signals for ADAS. In: Schulze, T., Müller, B., Meyer, G. (eds.) Advanced Microsystems for Automotive Applications 2016. LNM, pp. 125–138. Springer, Cham (2016). https://doi.org/10.1007/978-3-319-44766-7_11
23. Pandey, A.K.T., et al.: Empirical evaluation of machine learning algorithms based on EMG, ECG and GSR data to classify emotional states (2013)
24. Bradley, M.M., Lang, P.J.: Measuring emotion: behavior, feeling, and physiology. Cogn. Neurosci. Emot. **25**, 49–59 (2000)
25. Lisetti, C.L., Nasoz, F.: Using noninvasive wearable computers to recognize human emotions from physiological signals. EURASIP J. Adv. Sig. Process. **2004**, 929414 (2004)
26. Santos Sierra, A., Ávila, C.S., Casanova, J.G., Pozo, G.B.: A stress-detection system based on physiological signals and fuzzy logic. IEEE Trans. Ind. Electron. **58**, 4857–4865 (2011)
27. Shalini, T.B., Vanitha, L.: Emotion detection in human beings using ECG signals. Int. J. Eng. Trends Technol. (IJETT) **4**, 3113–3120 (2013)
28. Tan, S., Guo, A., Ma, J., Ren, S.: Personal affective trait computing using multiple data sources. In: 2019 International Conference on Internet of Things (iThings) and IEEE Green Computing and Communications (GreenCom) and IEEE Cyber, Physical and Social Computing (CPSCom) and IEEE Smart Data (SmartData) (2019)
29. Berntson, G.G., et al.: Heart rate variability: origins, methods, and interpretive caveats. Psychophysiology **34**, 623–648 (1997)
30. Albraikan, A., Tobón, D.P., El Saddik, A.: Toward user-independent emotion recognition using physiological signals. IEEE Sens. J. **19**, 8402–8412 (2018)
31. Can, Y.S., Arnrich, B., Ersoy, C.: Stress detection in daily life scenarios using smart phones and wearable sensors: a survey. J. Biomed. Inform. 103139 (2019)

32. Greene, S., Thapliyal, H., Caban-Holt, A.: A survey of affective computing for stress detection: evaluating technologies in stress detection for better health. IEEE Consum. Electron. Mag. **5**, 44–56 (2016)

33. Wampfler, R., Klingler, S., Solenthaler, B., Schinazi, V., Gross, M.: Affective state prediction in a mobile setting using wearable biometric sensors and stylus. In: Proceedings of the 12th International Conference on Educational Data Mining (EDM 2019) (2019)

34. Bach, D.R.: A head-to-head comparison of SCRalyze and Ledalab, two model-based methods for skin conductance analysis. Biol. Psychol. **103**, 63–68 (2014)

35. Furuichi, K., Worsley, M.: Using physiological responses to capture unique idea creation in team collaborations. In: Companion of the 2018 ACM Conference on Computer Supported Cooperative Work and Social Computing (2018)

36. Kelsey, M.: Applications of sparse recovery and dictionary learning towards analysis of electrodermal activity (2017)

37. Reutermann, J.E., Traupe, O., Hedderich, J., Kaernbach, C., Stephani, U.: Sympathetic activity of PPR-positive adolescents: clinical study. Neuropediatrics **47**, P07–P18 (2016)

38. Benedek, M., Kaernbach, C.: Decomposition of skin conductance data by means of nonnegative deconvolution. Psychophysiology **47**, 647–658 (2010b)

39. Schwarz, G., et al.: Estimating the dimension of a model. Ann. Stat. **6**, 461–464 (1978)

40. Deng, Y., Chang, L., Yang, M., Huo, M., Zhou, R.: Gender differences in emotional response: inconsistency between experience and expressivity. PloS ONE **11** (2016)

41. Rohrmann, S., Hopp, H., Quirin, M.: Gender differences in psychophysiological responses to disgust. J. Psychophysiol. **22**, 65–75 (2008)

42. Alexander, M.G., Wood, W.: Women, men, and positive emotions: a social role interpretation. In: Gender and Emotion: Social Psychological Perspectives, pp. 189–210 (2000)

43. Singh, S.: Video on Demand (VoD) Market worth $87.1 billion by 2024, January 2020. https://www.marketsandmarkets.com/PressReleases/audio-video-on-demand-avod.asp. Accessed Feb 2020

«Man-Machine» Interaction: The Determinants of the Untact Service's Use

Arij Jmour[⊠]

Faculty of Economics and Management Science, Sfax, Tunisia
arijjmour@gmail.com

Abstract. The aim of this study is to expand Lee and Lee's (2019) research who suggests that a significant part of face-to-face services should be replaced by untact services in the future and that untact service is new and its theoretical bases still need to be developed. A major novelty of this study lies in the discovery of the «untact service» antecedents and the link between this service and the experience co-creation. Thus, companies must focus on individual consumers who seeking uniqueness and who are open to the technology as well as the convenience of this service. This makes it possible to fulfill the limits of the personal contact. However, this untact service depends on the type of product or service. The results have theoretical and managerial implications for untact services.

Keywords: Untact service · Experience co-creation · Convenience · Uniqueness · Social exchange theory

1 Introduction

The digital age 4.0 is characterized by unprecedented rapid changes in the business environment (Lee and Lee 2019). It has led to a change in consumer buying behavior, moving from traditional selling to digital channels, thanks to advanced digital technologies and easy access to the global market (Lee and Lee 2019). Advanced digital technologies are now a part of everyday life for most businesses and individuals (Lee and Lee 2019). Therefore, consumers are less dependent on the information provided by brands and suppliers (Lee and Han 2013; Horn et al. 2015; Amar et al. 2019; Kim et al. 2019; cited by Lee and Lee 2019). For a long time, the traditional concept of quality service by sellers was the minimum obligation to attract the attention of the customers. Consumers are nowadays demanding personalized services that meet their tastes, needs and lifestyles (Lee 2018b; Lee and Lim 2018; cited by Lee and Lee 2019). Only technology-based innovation systems can provide such personalized services (Lee and Lee 2019). More important, many tech-savvy consumers nowadays search for various sources of information on the web and know much more about the various aspects of the product or service they are interested in than the store vendors, especially the busy people who prefer to spend as little time as possible shopping without the intervention of store employees. In other countries such as the United States, contact personnel have disappeared (Lee and Lee 2019). Consequently, if consumers feel uncomfortable when interacting with other

© Springer Nature Switzerland AG 2020
M. A. Bach Tobji et al. (Eds.): ICDEc 2020, LNBIP 395, pp. 105–114, 2020.
https://doi.org/10.1007/978-3-030-64642-4_9

people, the sales model must switch from one operating mode to another to a model based on the intangible service which uses technologies to replace human aids (Lee 2018a and b; cited by Lee and Lee 2019). Among these technologies, we cite the "untact" service studied by Lee and Lee (2019) which facilitate meetings with consumers without direct contact with contact staff, or even in a non-face to face form. In fact, this technology replaces human contact with human-machine interactions.

Through a large literature review, we can advance two types of interactions such as interaction between people and interaction between the person and the automated agent. The first type concerns contact staff. The second includes the computer, the mobile phone and the automatic machines, the consumer of which uses them to interact with others or to buy without human intervention. Thus, the consumer is able to interact, currently, with the medium he uses and with the content that companies produce (Hoffman and Novak 1996; cited by Hainla 2015, p. 4). This is well proven by Chiou et al. (2019) noting the interactions between humans and the automated agent. They explained that people treat computers like social actors, and engage in social exchanges with computers through the reciprocity of which these automated agents imitate social exchanges between people. In this context, the theory of social exchange founded by Emerson (1976) considers interactions between agents as transactions or exchanges of material goods and non-material goods, such as symbols of approval or prestige. Since then, traditional sales have considerably decreased and many reputable companies have disappeared or found themselves in a difficult financial situation. The main causes of bankruptcy or financial difficulties of traditional stores include the impact of technological innovations such as standalone stores without staff. Fast online and mobile services, mobile applications, shops with smart robots, interactive kiosks and untact services have emerged. However, the untact service is new and its theoretical foundations still need to be developed (Lee and Lee 2019). Otherwise, although it is an innovative service, it is not rigorously supported by the literature due to the lack of knowledge accumulated on this new type of service (Lee and Lee 2019). Therefore, in their future research paths, Lee and Lee (2019) insisted that the untact service remains to be demonstrated by empirical research. An understanding by researchers of its effect on the consumer therefore seems fundamental today. The objectives of this study are (1) to examine the evolution of consumer-centered service and task transfer, (2) to explore through a qualitative study the antecedents which led to the emergence of "non-contact" as a new service strategy. The theoretical interest of this study is to analyze the opportunities and challenges related to the implementation of the "untact" service. This has important implications for companies wishing to have a competitive advantage in the digital age and encourage their customers to constantly seek the type of service they want.

2   Theoretical Framework

2.1   The Value

"Untact" is a keyword created in South Korea by adding the prefix "un", which has the meaning of "no", to the word "contact" (Kim et al. 2018; Lee and Lee 2019). Untact service refers to a service provided without face-to-face meetings between employees and consumers through the use of digital technologies. The Untact service is generally launched by the consumer who wants a service without engaging in a personal meeting (Fitzsimmons et al. 2008; Bordoloi et al. 2018).

2.2   The Value Co-creation

Technology-supported innovation can shift certain responsibilities related to transaction tasks from businesses to consumers. Task transfer is not new. Many companies have offered self-service and unmanned information kiosks. What is new in the digital age is that the degree of task transfer is greater than traditional dating. Thus, encouraging consumers to design and develop new products and services through co-creation is an area of major interest for researchers and marketing managers (Prahalad and Ramaswamy 2004a, b and c; Essamri et al. 2019; cited by Roy et al. 2019).

3   Methodology

Our interest is in the case of a smart automatic machine that offers consumers the possibility to choose and buy without any human interaction. The data that served as the basis for proposing the history of untact service's use was collected through a field survey carried out at the exit of a pastry shop located in Sousse, Tunisia, with a convenience sample made up of 13 interviewed until reaching semantic saturation. This pastry shop sells hot and cold drinks only through an automatic machine. The survey was administered face to face, outside the pastry shop. This sample consists of 9 women and 4 men of quite varied age and CSP (see Appendix 2). Some are used to using this machine, others are using it for the first time. Our sampling unit represents anyone who leaves the pastry shop using the automatic machine for hot and soft drinks and who voluntarily agrees to respond to our interview guide. For better internal validity, data collection took place throughout the week at different times of the day, excluding any exceptional or busy periods. The participants were questioned around six major themes (see Appendix 1). The semi-structured interviews were fully transcribed and a thematic analysis of the speeches was carried out (Table 1).

Table 1.   Topics of the interview guide

Topic 1	Reasons of using the automatic machine
Topic 2	All of the thoughts that the interviewees may have had and the emotions they may have felt during the experience of using the automatic machine
Topic 3	The opinion of interviewees concerning the use of the automatic machine without human intervention
Topic 4	Assessment of interviewee satisfaction compared to interaction with contact staff)
Topic 5	Intention of future behavior towards the use of this machine
Topic 6	Other types of smart dispensers and the reasons of their use

4 Determinants of Untact Services: Key Findings

The thematic analysis reveals the existence of several determinants of the "untact" service's use.

1. Convenience

Based on the analysis grid of occurrence frequencies (cf. Appendix 3), we were able to explore convenience as the most important determinant (38.5%) compared to the other determinants identified in the qualitative study. According to frequency calculations, rapidity (27%) was the most important proportion of convenience. It is followed by ease of use (11.5%). First, almost all interviewees cited the verbatim *«instead of waiting a long waiting line»*. In this context, Kim et al. (2018) clarified that consumers want to receive fast services at any time. Equally, Berry et al. (2002, p. 5) defined convenience as «the time and effort saved by consumers when purchasing and using a service» (cited by Roy et al. 2018, p. 294). Untact technology can therefore be considered as a technology that allows receiving quickly and on time (Lee 2018a and b, 2019). What is shown in the speeches of the interviewees: *«I take my need quickly and I go out, especially in the event of a crowd in the pastry shop»* (Interviewee 5, student), *«it's always fluid in the sense that you have not to wait, you don't have to say wait, I'm going to do you a favor or why you served the other before me»* (Interviewee 1, teacher), *«especially for someone engaged and overworked all the time»* (Interviewed 8, teacher). It is clear through these remarks that professionals or very busy households prefer to spend the least possible time to do their shopping without the intervention of store employees. Also, among the comments relating to the value of convenience, 11.5% of those interviewed announced the ease of use of this intelligent machine: *«I don't need to search too much, it makes me a lot easier.»* (Interviewee 5, student), *«It's a way that makes life easier* "(Interviewee 8, student). Thus, the "untact" service increases consumer comfort and their preferences for technology (Kim et al. 2018; cited by Lee and Lee 2019): *«I feel more comfortable»* (Interviewee 13, student). These technologies have significantly improved the dynamic capabilities of businesses, allowing agility, flexibility and adaptability to align business strategies with changes in the environment, particularly to meet consumer needs and preferences which evolve rapidly. So the ideas for 24/7 service have created numerous automatic machines, self-service counters and answers to frequently asked questions. Therefore, convenience of service as well as "do it yourself" emerged as an important marketing strategy.

2. Individualism and the search of uniqueness (18.2%)

As suggested by Kim and al. (2018) and Lee and Lee (2019), another main cause of the emergence of this type of service is the fact that many young consumers accustomed to digital devices tend to feel uncomfortable with people and prefer solo shopping. This trend of silent consumer service is growing and spreading rapidly. As a result, one-person consumers often focus on their personal experience. They seek uniqueness and demand personalized services that meet their tastes, needs and lifestyles. These consumers have different service preferences during the encounter, depending on their psychological characteristics. The following remarks illustrate the presence of this psychological character: «*I do not need to have human contact and in the mornings generally I do not want to speak... I just pressed a button to have my choice*» (Interviewee 1, teacher), «*she gives me my need without breaking my head with a saleswoman*» (Interviewee 3, student), «*I prefer to buy and hit buttons in a personal way*» (Interviewee 6, engineer), «*I focus more when I'm alone*» (Interviewee 7, housewife), «*I'm not obliged to tell the saleswoman to give me coffee*» (Interviewee 9, student).

3. Limits of the contact staff

In addition to the reasons mentioned for using a smart automatic machine, 17.3% of those interviewed told us that the limits of contact staff prompted them to adopt such a service: «*I find non-smiling salespeople until I eliminate my purchase*» (Interviewee 3, student), «*sometimes the seller can't hear me and don't hear my exact need*» (Interviewee 5, student), «*the staff are happy with working hours... I don't find this advantage with the seller ... This intelligent distributor is without human mistake*» (Interviewee 6, engineer), «*The saleswoman who follows me to tell me do you need something... it bothers me*» (Interviewee 7, housewife), «*she can force me to buy other products*» (Interviewee 13, student). For this type of consumer, interactions with machines can replace human contact. Thus, if consumers feel uncomfortable when interacting with other people, the sales model must switch from one operating mode to another to a model based on the intangible service which uses technologies to replace human aids (Lee 2018a and b). Therefore, untact technology can be seen as a form of digital transformation in which the service paradigm is shifted from interaction with the consumer to intangible interactions based on advanced technologies.

4. Utility

6.7% of consumers use this untact service for its usefulness: «*He sees the machine as a simple necessity, taste is not that, he classifies this taste as industrial, he prefers to hear the sound of a coffee maker*» (Interviewee 1, teacher), «*because the coffee from the intelligent machine is hotter whereas in a tea room the waiter gives it to me cold or burnt*» (Interviewee 7, housewife), «*Sometimes I withdraw more than once a day, imagine that I always wait for the waiting line*» (Interviewee 7, housewife).

5. Discovery and openness to technology

Interviews revealed that 5.8% of respondents are very open to new technologies. They attach importance in particular to technologies without human contact. This untact technology can meet the demands of consumers who feel more comfortable with digital channels than those with technology problems (Lee and Lee 2019). Companies must find ways to indirectly contact their customers who prefer not to be contacted by their employees (Lee and Lee 2019). Besides, the respondents declared that they want to discover this new technology: «*Since I saw it for the first time, I wanted to discover how to use it, it is the first time that I see a thing like that, I also wanted to know if his hot chocolate is delicious*» (Interviewee 4, student), «*I feel up to date*» (Interviewee 8, teacher). According to Wikipedia, openness to technology refers to the appreciation of new ideas, curiosity and imagination.

6. The type of product or service

5.8% of respondents conveyed to us the idea that the untact service differs according to the characteristics of the product or service (Lee and Lee 2019). As recommended by Lee and Lee (2019), these interviewees warned us that an analysis of the characteristics of the product or service must be undertaken before the effective application of technologies: «*on the contrary, intelligent distributors of food products are risky*» (Interviewee 2, student), «*For example in a drugstore I need the opinion of the saleswoman*» Interviewee 7, housewife), «*it depends on the product of course*» (Interviewee 8, teacher).

7. Familiarity with technology

4.8% of the interviewees are either familiar with the technology «*it's clear and clean*» (Interviewee 1, teacher), "*it gives me a guaranteed thing*" (Interviewee 2, student), «*I have always used it in Algeria, this cafeteria next door I do not know its coffee well while this machine I know it well*» (Interviewee 10, retired), either with the product: «*I like the taste of this industrial coffee*» (Interviewee 1, teacher)," I buy it several times it is delicious» (Interviewee 12, employee). As advocated by Lee and Lee (2019), consumers who are familiar with digital devices tend to prefer solo purchases over purchases based on interactions with staff. In addition, many tech-savvy consumers search for various sources of information on the web and often know much more about the various aspects of the product or service they are interested in than the friendly store vendors.

8. Standardization

2.9% of the interviewees are inclined to assume that the untact service offers them a certain standardization: «The product in the intelligent machine is standard» (Interviewee 4, student), «*The interactive kiosk has a single language of figures understandable by everyone world*» (Interviewee 6, engineer).

9. The co-creation of the experience

The company offers technology to the consumer, and the consumer in turn creates his own experience: «*I prefer to buy and press buttons in a personal way*» (Interviewee 6, engineer), «*I take my needs without no one forces me to do anything*» (Interviewee 13, student). In this context, Roy et al. (2019) have shown that consumers create memorable experiences using the latest smart technologies. Their results showed that smart service influences the co-creation of smart experiences. Among several technologies, smart technology is one of the most influential technologies to affect behavior in co-creation (Grewal et al. 2017; Inman and Nikolova 2017). Recently, service companies have taken advantage of the use of these technologies to facilitate their co-creation activity (Voorhees et al. 2017). In addition, these intelligent consumers collectively share and co-create experiences (Cova and Dalli 2009). Thus, companies must provide experiential services, which generate value through an enjoyable experience.

5 Conclusion

In modern society, ICT has been used as a basic technology in all aspects of life. These have influenced all sales (Lee 2019). This study enriches the work exploring the concept and evolution of the "untact" service as a new customer service strategy and provided conceptual insights on the determinants of the "untact" service. It offers a first theoretical perspective which exposes the determinants of the "untact" service and their consequences on the co-creation of the experience. This allows managers to better apply this service. The main limitation of this study is that the study focused on only one type of technology such as the automatic machine. So one avenue of research would be to focus on other types of smart services. Another limitation of this research is inherent in its qualitative methodology, which prohibits any generalizing aim of the results. It would be interesting to deepen these results by extending them through a quantitative study. Finally, a limit rests on the fact that the size of the sample (13 interviewed) and its non-representativeness compared to the whole of the Tunisian population. It seems necessary to duplicate this study with a larger sample in order to confirm the results. These elements should therefore be considered in future research.

Appendix 1: Characteristics of the Sample

Respondents	Gender	Age	Catégorie socioprofessionnelle	Town
Interviewed 1	Woman	31	University teacher	Sousse
Interviewed 2	Man	21	Student	Sousse
Interviewed 3	Woman	21	Student	Sousse
Interviewed 4	Woman	20	Student	Sousse
Interviewed 5	Woman	20	Student	Sousse
Interviewed 6	Man	29	Engineer	Sousse
Interviewed 7	Woman	47	Housewife	Sousse
Interviewed 8	Woman	28	University teacher	Sousse
Interviewed 9	Woman	21	Student	Sousse
Interviewed 10	Man	65	Retired	Sousse
Interviewed 11	Woman	32	Employee	Sousse
Interviewed 12	Man	24	Employee	Sousse
Interviewed 13	Woman	20	Student	Sousse

Appendix 2: Vertical and Horizontal Analysis of the Analysis Grid

Interview n°	1	2	3	4	5	6	7	8	9	10	11	12	13	Frequencies	
														Absolute	Relative
Determinants of untact service's use															
Convenience	1	3	0	0	1	0	1	4	1	0	1	0	0	12	11.5%
Rapidity	4	1	3	0	3	2	3	3	2	2	2	1	2	28	27%
Ease of use	5	4	3	0	4	2	4	7	3	2	3	1	2	40	38.5%
Total	4	0	1	1	1	1	3	1	3	0	0	2	2	19	18.2%
Individualism	1	0	1	1	2	3	0	1	2	1	1	2	3	18	17.3%
Limits of the contact staff	1	0	0	0	0	0	1	1	3	0	0	1	0	7	6.7%
Utility	1	0	0	3	1	0	0	1	0	0	0	0	0	6	5.8%
Openness to technology	1	1	0	1	0	0	2	1	0	0	0	0	0	6	5.8%
The type of product or service	1	1	0	0	0	0	1	0	0	2	0	1	0	5	4.8%
Familiaritu with technology	0	0	0	1	0	1	0	0	0	1	0	0	0	3	2.9%
Standardization	1	3	0	0	1	0	1	4	1	0	1	0	0	12	11.5%
Total of categories														104	100%

Appendix 3: The Untact Machine

References

Berry, L.L., Seiders, K., Grewal, D.: Understanding service convenience. J. Mark. **66**(3), 1–17 (2002)

Bordoloi, S., Fitzsimmons, J., Fitzsimmons, M.: Service Management: Operations, Strategy, Information Technology, 9th edn. McGraw-Hill, New York (2018)

Chiou, E.K., Lee, J.D., Su, T.: Negotiated and reciprocal exchange structures in human-agent cooperation. Comput. Hum. Behav. **90**, 288–297 (2019)

Cova, B., Dalli, D.: Working consumers: the next step in marketing theory? Mark. theory **9**(3), 315–339 (2009)

Emerson, R.M.: Social exchange theory. Annu. Rev. Sociol. **2**(1), 335–362 (1976)

Fitzsimmons, J.A., Fitzsimmons, M.J., Bordoloi, S.: Service Management: Operations, Strategy, Information Technology, p. 4. McGraw-Hill, New York (2008)

Grewal, D., Roggeveen, A.L., Nordfält, J.: The future of retailing. J. Retail. **93**(1), 1–6 (2017)

Hainla, L.: Content Marketing–the key to effective brand communication in digital media (2015)

Inman, J.J., Nikolova, H.: Shopper-facing retail technology: a retailer adoption decision framework incorporating shopper attitudes and privacy concerns. J. Retail. **93**(1), 7–28 (2017)

Lee, D.: Strategies for technology-driven service encounters for patient experience satisfaction in hospitals. Technol. Forecast. Soc. Change **137**, 118–127 (2018a)

Lee, D.: Effects of key value co-creation elements in the healthcare system: focusing on technology applications. Serv. Bus. **13**(2), 389–417 (2019)

Lee, S.M.: Innovation: from small "i" to large "I". Int. J. Qual. Innov. **4**(1), 2 (2018b)

Lee, S.M., Lee, D.: "Untact": a new customer service strategy in the digital age. Serv. Bus. **14**, 1–22 (2019)

Lee, S.M., Lim, S.: Living Innovation: From Value Creation to the Greater Good. Emerald Publishing Limited, Bingley (2018)

Lee, S.M., Olson, D.L.: Convergenomics: Strategic Innovation in the Convergence Era. Routledge, Abingdon (2016)

Lee, S.M., Lee, D., Kim, Y.S.: The quality management ecosystem for predictive maintenance in the Industry 4.0 era. Int. J. Qual. Innov. **5**(1), 4 (2019)

Lee, S.M., Olson, D.L., Trimi, S.: Strategic innovation in the convergence era. Int. J. Manag. Enterp. Dev. **9**(1), 1–12 (2010)

Lee, S.M., Ribeiro, D., Olson, D.L., Roig, S.: The importance of the activities of service business in the economy: welcome to the Service Business. Int. J. **1**, 1–5 (2007)

Lee, Y., Han, J.: The rise of single-person households and changes in consumption patterns. KIET Ind. Econ. Rev. **18**(4), 16–28 (2013)

Prahalad, C.K., Ramaswamy, V.: Co-creating unique value with customers. Strategy Leadersh. **32**(3), 4–9 (2004a)

Prahalad, C.K., Ramaswamy, V.: Co-creation experiences: the next practice in value creation. J. Interact. Mark. **18**(3), 5–14 (2004b)

Prahalad, C.K., Ramaswamy, V.: The Future of Competition: Co-creating Unique Value with Customers. Harvard Business Press, Boston (2004c)

Roy, S.K., Shekhar, V., Lassar, W.M., Chen, T.: Customer engagement behaviors: the role of service convenience, fairness and quality. J. Retail. Consum. Serv. **44**, 293–304 (2018)

Roy, S.K., Singh, G., Hope, M., Nguyen, B., Harrigan, P.: The rise of smart consumers: role of smart servicescape and smart consumer experience co-creation. J. Mark. Manag. **35**, 1–34 (2019)

Voorhees, C.M., et al.: Service encounters, experiences and the customer journey: defining the field and a call to expand our lens. J. Bus. Res. **79**, 269–280 (2017)

Investigating Travelers' Dilemma in Managing Their Online/Offline Presence on Travel Experience

Inès Mestaoui[1](✉) and Mourad Touzani[2]

[1] Higher Institute of Management of Tunis (PRISME), Tunis, Tunisia
Ines.mestaoui@gmail.com
[2] Neoma Business School, Campus de Rouen, Mont-Saint-Aignan, France
mourad.touzani@neoma-bs.fr

Abstract. Several researchers have studied the impact of new technologies on the travel experience and how their use has transformed it. We joined the conversation, which says that the (dis) connection behavior is a consequence of the use of mobile technologies during the travel experience. In this work, we focus on the dilemma experienced by the young Tunisian traveler when she/he uses mobile technologies and how she/he chooses to be present/absent during her/his travel experience. An exploratory study was conducted with young Tunisian travelers to try to understand their digital behavior during the travel experience. The major contribution of this work is to try to understand the dilemma of (dis) connection and to present how the traveler experiences it.

Keywords: Mobile technologies · Travel experience · Traveler behavior · (dis) connection

1 Introduction

The revolutionary impact of new technologies is undeniable. These innovations reinvent everyday life and provide a new and unique value to the latter. Individuals focus today on technological innovation that creates memorable experiences on an individual and community level (Tussyadiah and Zach 2011). It is undoubtedly during the last decade that ICTs have become faster and more efficient for experience progress (Neuhofer et al. 2012). The latter had been more sophisticated to integrate an increasingly central place in the user environment (Lamsfus et al. 2013). Today smartphones are no longer single-function devices used only for communication (Zheng and Ni 2010; Reiter 2014). They now incorporate the same functions as a computer connected to the Internet (Lepp 2014). Furthermore, mobile technologies are changing the practice and the experience of leisure (White and White 2007). They assist individuals in leisure activities (Lachance 2016). They make leisure more meaningful, allowing the inclusion of friends not physically present at the time of the experience. They facilitate riskier recreation and simplify travel to remote places (Lepp 2014). Individuals depend on their mobile technologies to access

© Springer Nature Switzerland AG 2020
M. A. Bach Tobji et al. (Eds.): ICDEc 2020, LNBIP 395, pp. 115–125, 2020.
https://doi.org/10.1007/978-3-030-64642-4_10

the Internet, mobile apps, and virtual communities for communication and information access (Neuhofer and Ladkin 2017). Mobile apps and online communities are accessible whenever the user needs them, and wherever they are required (Not and Venturini 2013).

Travel, among other leisure experiences, involves constant interactions between the traveler and mobile technologies (Hwang 2010). Mobile devices contribute to the development of the travel experience (Wang et al. 2011). They simplify and enhance the traveler mobility (Salvadore et al. 2015). Travel experience is today more effortless through mobile technologies rapidity of use and ease of access to information (Gretzel 2010). Travelers cherish their autonomy and the ability to manage their travel experience using mobile devices (Liu and Law 2013). They book hotels, buy their plane tickets using their smartphones (No and Kim 2015). They also enhanced the sharing and exchange between users in order to facilitate the construction of the travel experience (No and Kim 2015). Travelers when they are using mobile technologies have created new habits (Cousin and Réau 2009) relying on others feedbacks to organize their travel experiences (No and Kim 2015). The daily use of mobile technologies has indeed created new ambiguities for the travel experience (Lachance 2016). The decision to connect or to disconnect is no longer straightforward (Lachance 2016). Using mobile technologies leads to blurred lines between online and offline worlds (Šimůnková 2019). When using mobile technologies during travel experience, the traveler is present in her/his daily life and to her/his entourage, which hinders the essence of the travel experience that used to be in line with departure, absence, escape, and freedom (Jauréguiberry and Lachance 2017; Neuhofer and Ladkin 2017). Trying to conciliate the objective of her/his travel experience and her/his daily life obligations, the contemporary traveler finds herself/himself continually oscillating between the two worlds.

We, therefore, join the conversation that examines the (dis) connection habits of the contemporary traveler by focusing on the absence/presence ambivalence as experienced by young Tunisian travelers. Few studies have studied (dis) connection in the tourism sector (Neuhofer et al. 2014b; Neuhofer 2016; Jauréguiberry 2014; Lachance 2016; Jauréguiberry and Lachance 2017). However, no research has focused, to the best of our knowledge, on how young Tunisian technology aficionado travelers are managing their presence/absence during the travel experience. We will try to understand, in this paper, how they are managing their (dis) connection process when they use their mobile apps during the travel experience.

2 Literature Review

New technologies are at the origin of essential transformations of individuals' behaviors and lifestyles. Mobile technologies provide access to a large number of apps and services. The use of mobile technologies has changed planning and travel behaviors. Thanks to the autonomy conferred by these mobile devices, decisions related to the travel program are made in situ (Gretzel et al. 2006). Indeed, travelers use various mobile apps. Whether for accommodation, transport, geo-tracking, immediate search for information and advice, communication, translation, or for entertainment, mobile apps are playing today an ever-growing role in the travel experience (Buhalis and Foerste 2015). These apps facilitate the travel experience by adding utilitarian, hedonic, and social values

(Belk 2014; Im and Hancer 2014; Middleton et al. 2014; Wang et al. 2014b). They may also be considered as intrusive because they can distract travelers from essential and valuable social or authentic experiences (Turkle 2011; Przybylski et al. 2013). Instead of fully immersing in the experience and enjoying every moment spent elsewhere, some travelers find themselves glued to their devices, unable to miss even the smallest detail of what is happening in their home country.

The travel experience is now dependent on these mobile devices. The daily and continuous use of these devices has brought about an era of "*constant-connectivity*" (Schlachter et al. 2015). This connectivity has developed the contemporary travel experience (Neuhofer and Ladkin 2017). This latter is easy and accessible through mobile technologies use, which ensures the exchange and social interaction (Salvadore et al. 2015). Indeed, travelers organize their travel experiences, communicate with their entourage, and share their experiences, regardless of their physical location (Neuhofer et al. 2014a). However, this connectivity impacted the travel experience. The latter is no longer what it used to be (Wang et al. 2012; Wang et al. 2016). Travel experience used to allow the traveler to escape her/his daily life, forget about stress and obligations and live an '*out of time*' experience (Schlachter et al. 2015). Mobile technologies have changed the modalities of physical mobility (Salvadore et al. 2015). Ubiquity allows the traveler to access a "hybrid *space*" (Salvadore et al., 2015), allowing the traveler to navigate in an environment that is both physical and virtual (Rallet et àl. 2009; Badot and Lemoine 2013). The traveler automatically switches between the two worlds blurring the boundaries between the two online/offline (Šimůnková 2019). The latter are even qualified as non-existent (Šimůnková 2019). The overlap between the virtual and real-world during the travel experience complicated the behavior of the contemporary traveler (Cliquet and Dion 2002; Dion and Michaud-Trévinal 2004; Salvadore et al. 2015). The latter is torn between online and offline worlds during travel experience (Lachance 2014) as the experience is expanding to both worlds (Neuhofer et al. 2014a, 2014b).

The Tunisian cultural context is a particular one regarding the Internet and social media. Tunisia in the 2000s was described by reporters without borders as Internet enemy (Perkins 2014). It has also been recognized as one of the three most authoritative and restrictive countries in the world when it comes to Internet access by Forbes magazine. Tunisia has always refused citizens unlimited access to the Internet (Perkins 2014). In a context marked by censorship, no one would have believed in the role of the Internet and social media to support the Tunisian revolution (Ben Henda 2011). Since that, social media, and, more particularly, Facebook, play a crucial role in relaying political, economic, and social information (Allagui and Kuebler 2011). Accordingly, mobile technologies and social media have replaced the traditional mass media communication (Etling et al. 2014). They have been a significant windfall for Tunisians' autonomy and a tool to support and maintain individual freedoms (Benkler 2006). They also allow the daily information relay. Tunisians have acquired a reflex to systematically access social media as soon as they wake up to see what is new in their digital environment (Jamali 2014).

3   Methodology

To understand Tunisian travelers are managing their (dis) connection behavior when using mobile technologies when traveling, we conducted a qualitative research based on in-depth interviews with 18 young Tunisian travelers relying on the gender (9 Female and 9 Male travelers) and the educational level (see profiles in Appendix 1). We used the Tunisian lived travel experiences as the main source of information and *"empirical evidence"* (Thompson 1997; Arsel 2017, p. 3). We chose to focus our research on young Tunisian travelers, given the affinity the latter have with mobile technologies. Today in the post-revolutionary context, their usage is more than a mere Facebook use and rely on mobile technologies to organize and live their travel experiences. The objective is the richness of data (Patton 2002); we have favored travelers with different travel experiences. We interviewed young Tunisians who travel alone and organize their travel experiences autonomously. During the interviews, we adopted a conversational app-roach in order to put the informants at ease and to facilitate the exchanges. We fully transcribed all the interviews, and we did a thematic analysis under NVivo software. We used NVivo11 to code the keywords, units, sentences, or even paragraphs trying to bring out general themes (Bazeley and Richards 2000). We coded the data having the same significations. Coding makes it possible to understand the information collected and present it in a relevant way. The codes were then presented in themes to allow the analysis and interpretation of the results (Miles and Huberman 1994).

4   Results and Discussion

Our findings demonstrate that young Tunisians adopted mobile technologies and used them always on travel. According to the interviews conducted, mobile Apps' uses are very numerous: geo-tracking, translation, accommodation, work, transport, social inter-action… Connected technologies make every experience easier for the traveler. They are time saver. Travelers can carry on different activities without moving. With various apps at their disposal, travelers can contact people, search for information, locate themselves, find accommodation, or means of transport, and look for interesting events. Travelers do not depend anymore on others on-site when co-constructing their travel experience.

Globalization and digital technology promoted the emergence of generations of young individuals who are different and unique (Bouhdiba 2019). In Tunisia, social-ization processes have undergone profound transformations since the beginning of the 2000s. Young people today maintain extra-familial social networks, primarily based on new technologies and social media (WhatsApp, Facebook, and Twitter, etc.) (Bouhdiba 2019). Mobile technology illustrates the belonging to a community of social media (Jau-réguiberry and Lachance 2017). Young Tunisians are expressing the need to belong to groups, even virtual ones, and to maintain these relationships regularly. Tunisians reg-ularly consult their smartphones for information and to communicate with their family and friends. While traveling, *"old habits die hard,"* and the traveler is revived by the same curiosity and an unchanging desire for information and communication.

"…However, I am looking for WiFi access; it is mandatory, Internet access, it is mandatory, whether at the hotel or elsewhere, it is essential to keep in touch with

my family in Tunisia to see what is happening there, to connect…" (Interviewee 3, Female, 33 years old).

Moreover, the results allowed us to highlight that the young Tunisian traveler in her/his will to manage her/his online and offline presence during her/his travel experience goes through three phases of (dis) connection: The real disconnection where the traveler physically leaves her/his hometown. However, she/he remains emotionally connected to her/his entourage. The symbolic disconnection where beyond the real disconnection; the traveler begins to make disconnection attempts. Emotional disconnection where the traveler ends up separating herself/himself from mobile technologies and, therefore, from the other.

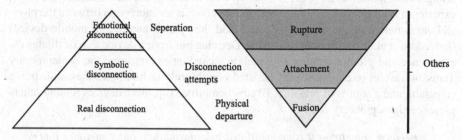

Fig. 1. The three phases of Tunisian traveler disconnection

- Fusion with others

The Real Disconnection: The real disconnection is experienced when the traveler leaves her/his country and flies to a destination far from her/his entourage. Connection is today the norm for all experiences, and more particularly for the travel one (Jauréguiberry 2014). Our findings demonstrate that ubiquity enhanced mobile technology use. The user may teleport herself/himself to her/his entourage whenever she/he wishes to do it. The traveler anchoring is no more than physical since the person is present in the destination, but she/he is experiencing cognitive and emotional mobility (Vergès 2005). The entourage can appear and disappear to the traveler through her/his (dis) connection. She/he may decide to invite her/his entourage whenever she/he share unique moments.

"…I did this in Bali, with … I had the connection access, and then I walked during the sunset in the rice fields it was so beautiful, and I found myself alone, so I wanted to share what I was living with my parents, and I shared that with my parents, I was there … live yes, they were contemplating with me…" (Interviewee 1, Female, 35 years old).

Mobile technologies are an emotional bond and a link to traveler daily life and her/his entourage. In this digital era, the real disconnection is less significant than the disconnection of mobile technologies and, therefore, of the other (Lachance 2014).

- Attachment to others

Symbolic Disconnection: Mobile technologies use increases the permeability between the online and offline world (Neuhofer and Ladkin 2017). Internet and social media may lead to information overload (Lee et al. 2017). The contemporary individual can no longer escape the news on social networks, professional emails, messages from those around him, and their calls (Christakis and Fowler 2009). Our findings demonstrate that continuous connection can resist to the desire for total escape and a complete change of scene inherent to the travel experience (Lachance 2016). Although attempts to disconnect are innumerable, they are not necessarily successful. The traveler is attached to her/his mobile devices (Jauréguiberry and Lachance 2017). She/he is continually brought back to the virtual space to keep up with what is going on in his life. The travel experience is therefore lived in a hybrid space, which is an interaction between the physical environment of the travel experience and the virtual interface of mobile devices (Salvadore et al. 2015; Šimůnková 2019). Deriving out from a device that facilitates the experience and promotes autonomy, mobile technologies converted the contemporary Tunisian traveler to a significantly assisted one. Mobile technologies may shift into a constraint and a source of pressure. Even when turned off, these devices remind others presence (Jauréguiberry 2014).

"… Because sometimes it is an email just for information, but sometimes it is an email that generates stress in a way that it requires an answer and if the person who sent it will activate the acknowledgment of receipt I will find myself obliged to answer, so it is for this reason that I do not access the emails, I cherish this disconnection to enjoy the trip…" (Interviewee 6, Male, 36 years old).

- Rupture with others

Emotional Disconnection: Contemporary travelers, manifest resistance to mobile technologies use (Jauréguiberry and Lachance 2017). Our findings show that Tunisian travelers try to disconnect seeking to manage their presence more efficiently in both worlds so that they can enjoy their travel experience intensely. Disconnection is mandatory to be able to travel (Lachance 2014). Contemporary traveler no longer wants to disperse themselves in these two worlds (Lachance 2014). They want to avoid distractions. They seek to be physically and emotionally present in the experience. They struggle to create a link with the destination and the various components of the experience environment to live their experience fully. They try to dissociate themselves from their virtual environment by turning off their phones (Jauréguiberry and Lachance 2017).

"…Me during my trip I forgot my phone, because all what I am looking for is freedom. I leave my daily life behind so I will not spend time looking at my phone or doing the same stuff that I usually do, I prefer to look up and down to contemplate, notice every new detail…" (Interviewee 13, Female, 25 years old).

The emotional disconnection cannot be experienced during the entire stay. In this continuous connectivity, it is indeed challenging to disconnect and take a moment for themselves (Lachance 2014). Emotional disconnection is necessary for freedom (Lachance 2014). The traveler has family responsibilities and commitments that make him always return to the online world (Lachance 2014; Neuhofer and Ladkin 2017). Disconnection during the travel experience is progressive at the beginning and barely punctual after that, but it is never definitive (Lachance 2014; Neuhofer and Ladkin 2017).

The dynamic of connection and disconnection while traveling is essentially decided concerning the other (Jauréguiberry and Lachance 2017). Social relationships have developed as a result of mobile technologies use. Contemporary individuals remain in continuous contact with families and friends they just left. The contemporary Tunisian traveler reinvents her/his relationships with others when she/he is using mobile technologies while traveling (Lachance 2016). This constant contact ensures "*social connectivity*" (Neuhofer and Ladkin 2017) by sharing travel experiences and interacting with others. The pleasure of sharing unique experiences with family and friends through social networks increases the pleasure deriving from the tourist experience (Munar and Jacobsen 2014).

Nevertheless, maintaining relations with their entourage can also affect the travel experience by creating and enhancing ambivalences between total immersion in the experience and maintaining communication and update of her/his life and entourage (White and White 2007; Neuhofer et al. 2014b; Wang et al. 2014b; Dickinson et al. 2016; Neuhofer and Ladkin 2017). Temporary interruption of these relationships is imperative in order to have an extraordinary travel experience (Lachance 2014). The traveler makes compromises with her/his entourage (Jauréguiberry 2014) to make regular contact with them and to reassure them just to be able to disconnect for the rest of the experience (Lachance 2014). Disconnection from the other is solely possible through these compromises (Jauréguiberry and Lachance 2017).

5 Conclusion

Today, contemporary individuals live under the almost absolute rule of new technologies (Hoffman and Novak 2018). Their excessive use affects travel experiences. The use of mobile Apps during travel experience provokes tensions related to travelers' behaviors (Lachance 2016). Today, connected travel takes place in a hybrid space that mixes presence and absence (Neuhofer and Ladkin 2017). The separation between the virtual and the real becomes ambiguous for tourists who permanently use mobile apps during their trip (Graillot 2005). This research aims to understand young Tunisian travelers' behaviors when they are using mobile apps during the travel experience, how they manage their online/offline presence, and how mobile apps impact the travel experience. The major contribution of this research is to point out travelers' dilemma online and offline worlds trying to manage their online presence to have successful travel experiences. Future research may focus on (dis) connection and how it impacts the form of the travel experience. Gender may also join the conversation, and we may study the particularities of (dis) connection of Tunisian women during travel experiences.

Results

Connection	Mobile technologies use	Geo-tracking
		Translation
		Accommodation
		Work
		Transport
		Social interaction
Disconnection	Real disconnection	The traveler physically separates herself/himself from the place she/he just left but remains emotionally connected to others
	Symbolic disconnection	More than physical separation, the traveler tries to make disconnection attempts
	Emotional disconnection	The traveler decides to separate from mobile technologies and therefore from others

Appendices

Appendix 1: Tables of profiles of interviewees

Interviewees	Gender	Age	Profession
Interviewee 1	Female	35	Teacher
Interviewee 2	Male	30	Employee at Emirates airlines
Interviewee 3	Female	33	Financial analyst
Interviewee 4	Female	30	Software engineer
Interviewee 5	Male	33	Textile engineer currently teaching English in Taipei
Interviewee 6	Male	36	Pharmacist
Interviewee 7	Male	26	Civil engineer
Interviewee 8	Male	26	Entrepreneur
Interviewees 9	Male & female	26 and 30 years old	Entrepreneurs
Interviewee 10	Female	32	Software engineer
Interviewee 11	Male	22	Student
Interviewee 12	Male	29	Employee in a company
Interviewee 13	Female	25	Consultant
Interviewee 14	Female	32	Manager in an insurance company
Interviewee 15	Male	30	Employee in a company
Interviewee 16	Female	29	Software engineer

(continued)

(*continued*)

Interviewees	Gender	Age	Profession
Interviewee 17	Female	26	University teacher
Interviewee 18	Male	27	Data analyst

References

Allagui, I., Kuebler, J.: The Arab spring and the role of ICTsI introduction. Int. J. Commun. **5**, 8 (2011)

Arsel, Z.: Asking questions with reflexive focus: a tutorial on designing and conducting interviews. J. Consum. Res. **44**(4), 939–948 (2017)

Badot, O., Lemoine, J.F.: Du paradigme dichotomique de l'expérience d'achat au paradigme ubiquitaire

Bazeley, P., Richards, L.: The NVivo Qualitative Project Book. Sage, Thousand Oaks (2000)

Belk, R.: You are what you can access: sharing and collaborative consumption online. J. Bus. Res. **67**(8), 1595–1600 (2014)

Ben Henda, M.: Internet dans la révolution tunisienne. Hermès, La revue **1**, 159–160 (2011)

Benkler, Y.: The Wealth of Networks: How Social Production Transforms Markets and Freedom. Yale University Press, London (2006)

Boddy, C R.: Sample size for qualitative research. Qual. Mark. Res. Int. J. (2016)

Bouhdiba, S.: Jeunes de Tunisie. Editions L'Harmattan (2019)

Buhalis, D., Foerste, M.: SoCoMo marketing for travel and tourism: empowering co-creation of value. J. Destination Mark. Manag. **4**(3), 151–161 (2015)

Christakis, N.A., Fowler, J.H.: Connected: The Surprising Power of Our Social Networks and How They Shape Our Lives. Little, Brown Spark (2009)

Cliquet, G., Dion, D.: Le comportement spatial du consommateur. Cliquet, Paris (2002)

Cousin, S., Réau, B.: Tourisme. EspacesTemps. net, **29** (2009)

Dickinson, J.E., Hibbert, J.F., Filimonau, V.: Mobile technology and the tourist experience: (dis)connection at the campsite. Tour. Manag. **57**, 193–201 (2016)

Dion, D., Michaud-Trévinal, A.: Les enjeux de la mobilité des consommateurs: de la gestion des stocks à la gestion des flux de clientèle. Décis. Mark. 17–27 (2004)

Etling, B., Kelly, J., Faris, R., Palfrey, J.: Mapping the Arabic blogosphere: politics, culture, and dissent. In: Media Evolution on the Eve of the Arab Spring, pp. 49–74. Palgrave Macmillan, New York (2014)

Graillot, L.: Réalités (ou apparences?) de l'hyperréalité: une application au cas du tourisme de loisirs. Recherche et Appl. Mark. (French Edn.) **20**(1), 43–63 (2005)

Gretzel, U.: Travel in the network: redirected gazes, ubiquitous connections and new frontiers. In: Post-Global Network and Everyday Life, pp. 41–58 (2010)

Gretzel, U., Fesenmaier, D.R., O'leary, J.T.: The transformation of consumer behaviour. In: Tourism Business Frontiers: Consumers, Products and Industry, pp. 9–18 (2006)

Hoffman, D.L., Novak, T.P.: Consumer and object experience in the internet of things: an assemblage theory approach. J. Consum. Res. **44**(6), 1178–1204 (2018)

Hwang, Y.H.: A theory of unplanned travel decisions: Implications for modeling on-the-go travelers. Inf. Technol. Tour. **12**(3), 283–296 (2010)

Im, J.Y., Hancer, M.: Shaping travelers' attitude toward travel mobile applications. J. Hospital. Tour. Technol. (2014)

Jamali, R.: Online Arab Spring: Social Media and Fundamental Change. Chandos Publishing, Hull (2014)

Jauréguiberry, F.: La déconnexion aux technologies de communication. Réseaux **4**, 15–49 (2014)

Jaureguiberry, F., Lachance, J.: Le voyageur hypermoderne. Érès (2017)

Kim, J., Tussyadiah, I.P.: Social networking and social support in tourism experience: the moderating role of online self-presentation strategies. J. Travel Tour. Mark. **30**(1–2), 78–92 (2013)

Lachance, J.: De la déconnexion partielle en voyage: l'émergence du voyageur hypermoderne. Reseaux **4**, 51–76 (2014)

Lachance, J.: Le smartphone: un objet transitionnel ou interactionnel? L'école des parents **6**, 105–114 (2016)

Lamsfus, C., Xiang, Z., Alzua-Sorzabal, A., Martín, D.: conceptualizing context in an intelligent mobile environment in travel and tourism. In: Cantoni, L., Xiang, Z(. (eds.) Information and Communication Technologies in Tourism 2013, pp. 1–11. Springer, Heidelberg (2013). https://doi.org/10.1007/978-3-642-36309-2_1

Lee, S.K., Lindsey, N.J., Kim, K.S.: The effects of news consumption via social media and news information overload on perceptions of journalistic norms and practices. Comput. Hum. Behav. **75**, 254–263 (2017)

Lepp, A.: The intersection of cell phone use and leisure: a call for research. J. Leisure Res. **46**(2), 218–225 (2014)

Liu, Y., Law, R.: The adoption of smartphone applications by airlines. In: Cantoni, L., Xiang, Z. (eds.) Information and Communication Technologies in Tourism 2013, pp. 47–57. Springer, Heidelberg (2013). https://doi.org/10.1007/978-3-642-36309-2_5

Middleton, C., Scheepers, R., Tuunainen, V.K.: When mobile is the norm: researching mobile information systems and mobility as post-adoption phenomena. Eur. J. Inf. Syst. **23**(5), 503–512 (2014)

Miles, M.B., Huberman, A.M.: Qualitative Data Analysis: An Expanded Sourcebook. Sage, Thousand Oaks (1994)

Munar, A.M., Jacobsen, J.K.S.: Motivations for sharing tourism experiences through social media. Tour. Manag. **43**, 46–54 (2014)

Neuhofer, B.: Value co-creation and co-destruction in connected tourist experiences. In: Inversini, A., Schegg, R. (eds.) Information and Communication Technologies in Tourism 2016, pp. 779–792. Springer, Heidelberg (2016). https://doi.org/10.1007/978-3-319-28231-2_56

Neuhofer, B., Buhalis, D., Ladkin, A.: Conceptualising technology enhanced destination experiences. J. Destination Mark. Manag. **1**(1–2), 36–46 (2012)

Neuhofer, B., Buhalis, D., Ladkin, A.: A typology of technology-enhanced tourism experiences. Int. J. Tour. Res. **16**(4), 340–350 (2014a)

Neuhofer, B., Buhalis, D., Ladkin, A.: Co-creation through technology: dimensions of social connectedness. In: Xiang, Z., Tussyadiah, I. (eds.) Information and Communication technologies in Tourism 2014, pp. 339–352. Springer, Vienna (2014b). https://doi.org/10.1007/978-3-319-28231-2_56

Neuhofer, B., Ladkin, A.: (Dis)connectivity in the travel context: setting an agenda for research. In: Schegg, R., Stangl, B. (eds.) Information and Communication Technologies in Tourism 2017, pp. 347–359. Springer, Cham (2017). https://doi.org/10.1007/978-3-319-51168-9_25

No, E., Kim, J.K.: Comparing the attributes of online tourism information sources. Comput. Hum. Behav. **50**, 564–575 (2015)

Not, E., Venturini, A.: Discovering functional requirements and usability problems for a mobile tourism guide through context-based log analysis. In: Cantoni, L., Xiang, Z. (eds.) Information and Communication Technologies in Tourism 2013, pp. 12–23. Springer, Heidelberg (2013). https://doi.org/10.1007/978-3-642-36309-2_2

Paris, C.M., Berger, E.A., Rubin, S., Casson, M.: Disconnected and unplugged: experiences of technology induced anxieties and tensions while traveling. In: Tussyadiah, I., Inversini, A. (eds.) Information and Communication Technologies in Tourism 2015, pp. 803–816. Springer, Lugano, Switzerland (2015)

Patton, M.Q.: Two decades of developments in qualitative inquiry: a personal, experiential perspective. Qual. Soc. Work **1**(3), 261–283 (2002)

Pearce, P.L., Gretzel, U.: Tourism in technology dead zones: documenting experiential dimensions. Int. J. Tour. Sci. **12**(2), 1–20 (2012)

Perkins, K.: A History of Modern Tunisia. Cambridge University Press, Cambridge (2014)

Przybylski, A.K., Murayama, K., DeHaan, C.R., Gladwell, V.: Motivational, emotional, and behavioral correlates of fear of missing out. Comput. Hum. Behav. **29**(4), 1841–1848 (2013)

Rallet, A., Aguilera, A., Guillot, C.: Diffusion des TIC et mobilité: permanence et renouvellement des problématiques de recherche. Flux **4**, 7–16 (2009)

Reiter, G.: Wireless connectivity for the internet of things. Europe **433**, 868MHz (2014)

Salvadore, M., Menvielle, L., Tournois, N.: Diffusion des services mobiles et mobilité du consommateur: une étude sur les déterminants et les conséquences des usages au cours d'un séjour touristique. Manag. Avenir **3**, 163–185 (2015)

Schlachter, S., McDowall, A., Cropley, M.: Staying "switched on" during non-work time: reviewing consequences for employees. In: Division of Occupational Psychology Annual Conference (2015)

Šimůnková, K.: Being hybrid: a conceptual update of consumer self and consumption due to online/offline hybridity. J. Mark. Manag. **35**(1–2), 40–74 (2019)

Tanti, A., Buhalis, D.: Connectivity and the consequences of being (dis)connected. In: Inversini, A., Schegg, R. (eds.) Information and Communication Technologies in Tourism 2016, pp. 31–44. Springer, Cham (2016). https://doi.org/10.1007/978-3-319-28231-2_3

Thompson, C.J.: Interpreting consumers: a hermeneutical framework for deriving marketing insights from the texts of consumers' consumption stories. J. Mark. Res. **34**(4), 438–455 (1997)

Turkle, S.: Life on the Screen. Simon and Schuster, New York (2011)

Tussyadiah, I., Zach, F.: The influence of technology on geographic cognition and tourism experience. In: ENTER, pp. 279–291, January 2011

Vergès, E.: Le paradoxe de la mobilité à l'heure du numérique et des réseaux: entre vitesse et cloisonnement. La pensee de midi **1**, 127–129 (2005)

Wang, D., Park, S., Fesenmaier, D.: An examination of information services and smartphone applications (2011)

Wang, D., Park, S., Fesenmaier, D.R.: The role of smartphones in mediating the touristic experience. J. Travel Res. **51**(4), 371–387 (2012)

Wang, D., Xiang, Z., Fesenmaier, D.R.: Adapting to the mobile world: a model of smartphone use. Ann. Tour. Res. **48**, 11–26 (2014a)

Wang, D., Xiang, Z., Fesenmaier, D.R.: Smartphone use in everyday life and travel. J. Travel Res. **55**(1), 52–63 (2014b)

Wang, D., Xiang, Z., Fesenmaier, D.R.: Smartphone use in everyday life and travel. J. Travel Res. **55**(1), 52–63 (2016)

White, N.R., White, P.B.: Home and away: tourists in a connected world. Ann. Tour. Res. **34**(1), 88–104 (2007)

Zheng, P., Ni, L.: Smart Phone and Next Generation Mobile Computing. Elsevier, Amsterdam (2010)

Interactivity: A Key Element for Future Digital Communication

Yoldes Rannen[1(✉)], Norchene Mouelhi Ben Dahmane[1,3], and Faten Malek[2]

[1] Institut des Hautes Etudes Commerciales Carthage, Université Tunis Carthage, Tunis, Tunisia
rannenyoldes95@gmail.com
[2] ESSCA School of Management, Angers, France
[3] Univ. Manouba, ISCAE, LIGUE LR99ES24, Campus Universitaire Manouba,
2010 Manouba, Tunisie

Abstract. This study explores the concept of interactivity and investigates its impact on the advertising field through the discovery of interactive online advertisements as a new communication style in the digital era, its technical feasibility and its attractiveness for both advertisers and consumers. The exploration of this concept was conducted through a qualitative exploratory research; a mini focus group with developers in order to investigate the technical feasibility of an interactive online advertisement and then semi-structured interviews with advertising experts and consumers to highlight their perception towards interactivity and its impact on communication. The study had demonstrated that interactive advertisements are feasible; it also showcased that both consumers and advertisers have a positive perception of online interactive advertisement and that it had a positive impact on the consumer's behaviour and perception towards advertisements.

1 Introduction

Advertising has a very important role in our everyday life because of its huge exposure and reach which became a key element for the consumer and has impacts on his behaviour, his experiences and his opinions. Advertisements nowadays are a determinant of image and way of life; they have a huge impact on the consumer's thinking as well as on the attitude towards themselves, the world around them (Frolova 2014; Daugherty et al. 2019).

Ever since its appearance, the advertising field faced multiple changes and went through diverse milestones that made it the way it is today. From that very first act of advertisement back in prehistoric times of ancient Egypt to 2000 BC a lot has changed. And as the time passed the format, the content, the purpose and the placements of advertisements kept on evolving until the appearance of online advertising, mobile advertising and many more.

Advertising has a huge impact on the consumer's behaviour as it was the reason for the appearance of personalisation and the importance of gathering information about the consumers to better relate and to ensure its effectiveness. The advertising field is characterized by its continuous evolution and growth. Urbanization and communication advancements had a huge impact on the facilitation of use of advertising, it is what

© Springer Nature Switzerland AG 2020
M. A. Bach Tobji et al. (Eds.): ICDEc 2020, LNBIP 395, pp. 126–138, 2020.
https://doi.org/10.1007/978-3-030-64642-4_11

explains the power and position it has nowadays on societies both on the economic and the cultural aspects (Frolova 2014).

The advertising field is being studied by numerous researchers since it is a vast and interesting subject that can have various facets and can be perceived differently depending on the used lenses. The growing interest in advertisements is not only because of its positive impact and position, but it is also because of the reaction of consumers towards it. Advertisements nowadays are hated by consumers which caused the appearance of multiple phenomena such as ad annoyance, ad avoidance and ad block. Multiple studies emerged to investigate the reasons for this change of perception and to study how to reduce this negative impact. The consumer behaviour changed and modern users hate ads because of their intrusiveness, lack of creativity and high frequency. It has emerged that customers are not only searching for representativity and personalization but they also want to engage with their entourage and feel empowered (Belanche et al. 2019).

With this being said, the concept of interactive content kept on evolving and gaining the interest of individuals. It changed from choices in a play during 1992 to the production of an entire Netflix movie with a storyline that evolves with the viewer's choices. This concept gained a lot of interest from consumers[1] which highlighted their need for interactivity and pushed major production companies such as Netflix and YouTube to opt for the interactive and engaging content to have better reach and effectiveness.[2]

The fact that interactive content has a very positive impact on the consumer behaviour and makes the viewers feel valorised and have a certain degree of control made us wonder if maybe the usage of such a content in advertising can change the consumer's perception of advertisements and their reaction towards them and these findings led us to the main question of the current study: what is an interactive online advertisement and is it the future of the advertising field as a key element for future digital communication which leads us to the following research questions:

1) How to create an interactive online advertisement?
2) What are the perceptions of advertising experts and consumers toward interactive online advertisement?
3) How the interactive online advertising impact consumer behaviour?

2 Literature Review

Interactive online advertising is a concept with few studies treating it that is why it was crucial to explore it through the combination of two existing and well treated concepts which are online advertising and interactive advertising.

2.1 Online Advertisements

Online advertising has become the most dominant and performing advertising mode; it is the best way to reach societies and individuals. Online advertising also called

---

[1] http://www.allocine.fr/article/fichearticle_gen_carticle=18679753.html.
[2] https://www.engadget.com/2019/10/23/youtube-a-heist-with-markiplier-trailer/.

internet advertising or web advertisement and even digital advertising is equivalent to advertisement through the online medium which is the internet (Einsend 2019).

The business dictionary defines online advertising as the *"Use of internet as an advertising medium where promotional messages appear on a computer screen. Since the communication software (browser) reveals sufficient information about the site's visitors, online advertising can be custom-tailored to match user preferences."*

Online advertising has evolved throughout the years; the market and the consumer faced a lot of changes which made it necessary to find new ways of advertising and to understand the evolution of the consumer's perception (Ha 2008; Ngowi 2015). And with the appearance of internet and the huge growth of the available consumer data, advertisers opted for data-driven advertising as their ideal online technique because it allows them to target a wider range of customers and to tailor the advertisement following their preferences through the processing, analysis and packaging of the collected data (Turrow 2017). The huge of amount of data and the negative perception of consumers towards advertisements made the advertiser obliged to come up with solutions in order to face ad avoidance and ad blocking and to reduce ad annoyance and hate (Goldstein et al. 2014; Söllner and Dost 2019).

2.2 Interactive Advertisements

Interactive advertising is one of the newest and most evolving types of advertising since it allows better communication with the consumer and it ensures that they have more power over the flow of the advertisement (Gromuls 2011; Ratchford 2015). Interactive marketing can be defined as the usage of a two-way communication channels in order to improve user contact and to allow a direct contact with the enterprise (Michalski 2017). Researchers showed an immense interest in interactive advertisement which can be highlighted by the numerous publications treating it and its evolution throughout marketing research history from the discovery of the concept to the study of it strategies and their impact on consumer behaviour.

In this consumer-centric market, understanding the consumer behaviour is no longer a choice; having a clear vision on the customer, his persona and all his characteristics is a must to the advertiser in order to know what to tell his consumer and how the message should be delivered; customer behaviour can be defined as *"the study of individuals, groups, or organizations and the processes they use to select, secure, and dispose of products, services, experiences, or ideas to satisfy needs and the impacts that these processes have on the consumer and society"* (Kuester 2012). The appearance and evolution of social media platforms have made the market evolve faster and wider and it made consumers have more power, knowledge, sophistication and drive which explain the frequent consumer behaviour change. Studies proved that nowadays the consumer's interest is no longer personalization or relevance; 21st century consumers actually want to create value within the advertisement and they want to conduct a real time interaction. Consumers no longer want to be on the end of the receiving line but they want to take part in the process of the creation (Rodgers and Thorson 2017); they want to be empowered and they want to have a certain degree of control over the advertisement and their experience while visualizing it (Belanche et al. 2019; Busser and Shulga 2019).

3   Methodology

This research aims to explore the concept of interactive online advertisements and to check the importance of interactivity as an element in digital communication and study its different sides that is why we opted for an exploratory qualitative research which will be conducted through three different steps; a mini focus group with software developers in order to investigate the feasibility of interactive advertisements and semi-structured interviews with both advertising experts and consumers; in order to study the attractiveness of the advertisement and the perception of the consumer for this new type of advertising. The individual interviews were accompanied by a video elicitation with the purpose to allow both samples a better understanding of the interactive online advertisement.

The method of triangulation was used in order to treat and compare the results obtained in diverse research techniques used and all data collected was analysed manually throughout thematic analysis.

3.1   Mini Focus Group with Developers

The creation of the interactive online advertising needs algorithms that will actually allow the interactivity and the creation of value by the user through the choices that will change the evolution of the story of the advertisement. In order to study its feasibility, we conducted a mini focus group with 5 developers who have different levels of expertise but work in the same company in order to be sure that they are exposed to the same work environment, culture and conditions. The sample of the focus group was composed of males and one female and their age ranged from 22 to 30 years old.

The mini focus group with developers was conducted using an animation guide (Appendix 1), at the participants' place of work and it was realized to figure out if the creation of such an advertisement is actually feasible and to investigate the convenient conditions to create it and the most eligible profiles to do so.

3.2   Semi-structured Interviews

As for the study of the attractiveness of the interactive online advertising semi structured interviews were conducted with both advertising experts and consumers. The sample of experts was created through the snowball technique; while the sample of consumers was defined through the quota technique because of the need of a wide group with different profiles and behaviours. The collection of data stopped when we reached semantic saturation; after 5 interviews with advertising experts and 19 interviews with consumers, there was no need to add more respondents since there was repetition in their statements and there were no new content.

- **Semi-structured Interviews with Advertising Experts**
  Advertising experts are a crucial sample in this study since the thought of launching a new format of advertisements can't be possible without studying the perception and the reaction of the experts who will be working with it. The sample was composed of 5 advertising experts; 4 males and 1 female their ages varying from 26 to 43

years old and they are working in different companies and agencies and with a proven knowledge and experience in the advertising field. Interviews with advertising experts lasted between 14 and 31 min and their main focus was the study their perception and reaction to this new advertisement while comparing it with the advertisements they use nowadays (Appendix 2).

- **Semi-structured Interviews with Consumers**
  To fully understand the concept and be sure that this format is attractive enough that consumers will watch it, there was a definite need to conduct semi-structured interviews to know their perception and opinions. Interviews were conducted with 19 digital users which mean that they just needed to be individuals who surf online and are actually exposed to online advertisements. The sample of consumers was composed by 10 females and 9 males; their age varied from 21 to 54 years old and they belonged to different socio-economic categories in order to ensure heterogeneity and diversity. The interviews lasted between 26 and 47 min, they allowed us to understand the consumer's perception and reaction to online advertisements to which he is exposed daily and to investigate his reaction towards the interactive online advertisement and study his opinion and behaviour (Appendix 3).

3.3   Video Elicitation

In order to be sure that the experts and the consumers studied are well aware of the attractiveness of the interactive online advertisement and that they have a clear idea about this new format, it was fruitful to opt for visual methods. That is why we created an interactive advertisement to allow both our targets to try out the advertisement themselves and check how it worked. That exposure meant to allow the respondent to check the new format and to have a better understanding and a clear perception about the whole concept in general. In order to create an interactive online advertisement two major elements needed to be defined: the content of the advertisement and the algorithm that will allow the interactivity throughout the advertisement. Once the development process was over and the carcass of the advertisement was ready, it was necessary to define the scenario of the advertisement and the definition of the diverse choices and options it would contain. This projective technique allowed us to study the impact on interactive online advertisement and to study in real time the reaction of the viewers whether the advertisers or the consumers.

4   Findings

Methodology	Findings
Mini focus group with developers	The study of the feasibility of the interactive online advertisement highlighted that it can be put in both web sites and social media platforms placements it just depends of the level of expertise of the developer and it can be in whatever format (video/image /text/hyperlink) because the created algorithm doesn't limit its input
	The duration and the cost varies depending on the developer and his competencies; junior developers needed a minimum of 6 months while the senior developer needs between 2 weeks to 3 months as for the price it varied depending on the duration the creation of the advertisement will take. They all stated that there is no need to preparations in advance
	As for the eligibility of the developer each one of them evoked his own speciality and the one he is more lenient to use but they all agreed that Python is the best choice someone can opt to use because of its multiple functionalities. *"A senior developer may be the best profile to get to start this because of the knowledge and the experience or a team of developers who know front and back development and who are leads it means they managed a team; as for the used language I am sure python is the best choice because of its multiple benefits"*
Individual semi-structured interviews with advertising experts	Experts defined an effective advertisement as the one that attracts customers the most and that has an efficient targeting. Most advertisers are not satisfied with nowadays online advertisements and highlighted that a changes needed to be conducted either the format, the frequency the'duration or the content. *"Reaches the target that is actually relevant to the product or service and has a purpose it needs to be intriguing and interesting in order to capture the interest of the target"*

(continued)

(*continued*)

Methodology	Findings
	Most respondents had a positive perception of the interactive advertisement. They enjoyed it and found it very interesting to use and adequate to nowadays consumers. While One expert didn't like the new format and didn't find it relevant to the Tunisian target
	Advertisers answers showcased that interactive online advertisements can be the future of the advertising field since they are feasible to all types of products and services and can work with all pricing models.
Individual semi-structured interviews with consumers	Consumers shared that ad characteristics such as format and placement influence their perception of online advertisements which showed that they are annoyed of ads and that is why they don't remember them and they tend on use blocking apps to avoid them.
	Online advertisements have direct impact o the consumers' brand perception and relation with the advertisement itself which is noticed through the changes they wish to conduct to advertisements such as delivery , duration, content, format, placement and frequency
	Consumers had a positive perception to interactivity; they enjoyed the interactive online advertisement and had a positive reaction to it; they were mostly impressed by the format, the creativity and how the content was presented. *"I find it very interesting because it is not intrusive and the viewer is the one going for it"*
	Interactive content was well received by consumers which was highlighted by their will to rewatch it and share it, they shared that it has a wide scope and that it will empower them and capture their curiosity; which highlights that interactive online advertising is attractive to consumer

5 Discussions

This section presents the findings of the conducted research and confronts them with past studies' findings in order to test the relevance of the research and to have a better understanding of the studied concepts.

5.1 Perception of Online Advertisements

The results show that the perception of online advertisements can be determined through the advertisement's characteristics, recall, conception and annoyance. It showcased that both consumers and advertising experts are not satisfied with the advertisements and that they need an upgrade. Consumers considered advertisements as intrusive and as an annoyance with no added value. These results match with the studies of Ghose et al. (2017) and Goldfarb and Tucker (2012) that exhibited that users consider advertisements as a source of annoyance and tend to avoid them because they are intrusive; they disrupt their user experience and interfere in their enjoyment process. The analysis of the results also showed that the perception of the advertisement is directly related to the format and it highlighted that its effectiveness improves with the usage of videos and music. The past studies showcased that visual and moving elements have a much bigger effect on the consumer, his concentration span and the advertisement recall (Burns and Lutz 2006).

5.2 Consumers' Behaviour Towards Online Advertisements

Previous marketing literature indicated that online advertisements have a negative impact on consumers' behaviour and that they caused the appearance of the phenomenon of ad avoidance and ad block (Speck and Elliott 1997; Wielki et al. 2018; Tudoran 2019) which is confirmed by our results that showcased that all consumers use ad blocking tools in order to reduce their exposure. Previous studies also highlighted the impact of advertisements on brand perception and its direct effect in the development of implicit attitudes towards the brand such as brand hate and preferences (Keller 2010; Madhavaram and Appan2010) which was also proved by our results.

5.3 Perception of Interactive Online Advertisements

The perception of interactive online advertisements was studied throughout the study of advertisers and consumers' reaction to the interactive advertisement. Results show that they are impressed by the new format and they perceive it as an innovative way to make the consumer more involved and engaged. The findings prove that their perception is positive which aligns with the previous marketing literature that highlights the importance of interactive advertisements and their potential to evolve and draw the attention of users (Rodgers and Thorson 2017; Park 2019).

5.4 Impact of Interactivity on Consumers' Behaviour

The analysis of the results showcase that interactive online advertisements have a huge impact on consumers' behaviour; they showed proof of attractiveness to the interactive advertisement and expressed their growing interest by highlighting their will to rewatch and share it. Advertising experts also indicated that it has a positive impact because of the growing need for interactivity and engagement in the process of value creation. They shared that this format can be the solution to reduce ad avoidance and block caused by the intrusiveness and the irrelevance of advertisements. Interactivity solves the problem of the short concentration span of the consumers by truly involving him in the creation process. These findings confirm those in previous studies that state that consumers have changed and that they are searching for interactivity because they no longer want to be passive elements in the process of value creation (Langaro et al. 2015). Marketing literature dug into the consumer behaviour change and the need of interactivity, studies showcased that his attitude toward the ad can change if he feels empowered and engaged which is why interactivity is crucial to give him control over the ad and their experience while visualizing it (Rodgers and Thorson 2017; Belanche et al. 2017; Belanche et al. 2019).

5.5 Vision of Interactivity

Marketing literature stated the need for a new format of advertisements that is creative and ensures the effectiveness of ads and gets over the rejection of consumers (Cheng et al. 2019; Liu-Thompkins 2019) and that the inclusion of interactivity will have a positive impact. Findings also highlighted the fact of the evolution of technologies and the huge amount of data which actually gave birth to data-driven advertising (Jiang et al.2017; Li 2019; Yu et al. 2019). The findings of previous researches actually are aligned with the vision of experts who shared that interactivity has the power to capture the curiosity of the viewer and it will keep on growing because of the evolution of technologies which will add to its scope.

6 Conclusion

The results of this study showcase its visible contributions in theory and in management. It allowed the exploration of a whole new concept which is interactive online advertisements by merging online advertising and interactive advertising, its technical feasibility and its attractiveness while studying consumers and advertising experts' perceptions. It paved the way to a discovery of a new effective format of advertising that has a positive effect on the consumers' ad perception, reduces the hate feeling and stops ad avoidance and ad block. It also highlighted that interactive content ensures a better reach and facilitates ad recall and brand awareness through the empowerment of consumers which suggests that maybe this new format is the solution to the lack of interest and the negative perception of modern consumers. Interactive content it catches the attention and the curiosity of the consumers which actually pushes them to finish the ad until the end in order to check the final output which enhances ad reception and reaction towards it. The fact that the consumer is the one engaging and creating value creates a connection between him and the ad so it eventually improves their perception of advertisements.

The limitations of this study suggest opportunities for future research. First, the scope of the research was limited since all participants were from Tunis, there were no significant representation of multiple generations and there was a huge gender inequality within the participants. We can also state as a limit the lack of measures due to lack of time we didn't conduct a quantitative research which would have been interesting in order to have quantifiable measures to the studied variables it would be worthwhile to conduct a quantitative research as the next step that will allow us to present different measurements to the variables and to integrate artificial intelligence in the process of the selection of the interactive online advertisement.

Appendix

Appendix 1: Mini Focus Group Animation Guide

Theme 0
Introduce yourself
Theme 1: Factors of the ad feasibility
1) What are the placements that can accept such an ad? And why?
2) What are the formats that can be used for the creation of such an ad? And why?
Theme 2: Developers' Perception of the ads' needs
3) How much time will the creation of such an ad take?
4) How much will the creation of such an ad cost?
5) What does it require as a preparation in advance?
Theme 3: Developers' eligibility
6) What are the needed qualifications to be able to create such an ad?
7) Which language will be used to develop such an ad?

Appendix 2: Advertising Experts Interview Guide

Theme 0
1) Introduce yourself
2) How would you describe ads nowadays
Theme 1: Experts advertising strategy
3) What is an effective advertisement?
4) What are the most important elements for a successful Ad?
5) What are the things that need to change in ads nowadays?
6) What are the elements of the ad that can block the engagement of the consumer?
7) What are the platforms and the formats used in your social media advertising campaign?
8) Which pricing model do you use?
Theme 2: Advertising experts perception of interactive online advertising
9) What do you think about such an ad?
10) It is feasible for different products/services?
11) Is it feasible for which purpose of advertisement?
12) Which model is used for the pricing of the ad?
13) What are the advantages and the disadvantages of such an ad?

Appendix 3: Consumers Interview Guide

Theme 0
1) Introduce yourself
2) Do you use Internet? When and for which purpose?
3) Do you watch advertisements?
4) How many times do you find yourself exposed to advertisements?
5) Where do you mostly watch ads?
6) What do you think about advertisements?
Theme 1: Consumers' perception of online advertisements
7) What are the most important elements of an ad?
8) What is the element that attracts you in an ad?
9) Do you remember any ads you have watched? If yes which one and why?
10) Does the format of the ad affect your judgement of an ad?
11) Does the placement of the ad affect your judgement of an ad?
12) Does the duration of the advertisement affect your judgement?
13) Does the position of the advertisement affect your judgment of the ad?
14) What do you think about the frequency of the ads in digital platforms?
15) Do you use ad blocking apps? Why?
16) What annoys you the most about the ads nowadays?
Theme 2: Impact of online advertisements on consumer behaviour
17) How likely would you watch the ad until the end?
18) When you enjoy watching an ad do you share it or rewatch it?
19) Does the advertisement affect your judgement of the product/service?
20) What would you change in nowadays ads?
Theme 3: Consumers' perception of interactive online advertisement
21) What do you think about this new format of ads?
22) What did you like?
23) What didn't you like about it?
Theme 4: Effect of interactive content
24) Will you rewatch it again? Why?
25) Will you share it with your friends? Why?
26) Do you think this type of ads can go with which products or services?
27) How do you see the future of interactive online advertisements

References

Belanche, D., Flavián, C., Pérez-Rueda, A.: User adaptation to interactive advertising formats: the effect of previous exposure, habit and time urgency on ad skipping behaviors. Telematics Inform. **34**(7), 961–972 (2017)

Belanche, D., Flavián, C., Pérez-Rueda, A.: Consumer empowerment in interactive advertising and eWOM consequences: the PITRE model. J. Mark. Commun. **26**(1), 1–20 (2019)

Braverman, S.: Global review of data-driven marketing and advertising. J. Direct Data Digital Mark. Pract. **16**(3), 181–183 (2015)

Busser, J.A., Shulga, L.V.: Involvement in consumer-generated advertising: effects of organizational transparency and brand authenticity on loyalty and trust. Int. J. Contemp. Hosp. Manage. **31**(4), 1763–1784 (2019)

Calder, B.J., Malthouse, E.C., Schaedel, U.: An experimental study of the relationship between online engagement and advertising effectiveness. J. Interact. Mark. **23**(4), 321–331 (2009)

Cheng, J.M.S., Blankson, C., Wang, E.S.T., Chen, L.S.L.: Consumer attitudes and interactive digital advertising. Int. J. Advert. **28**(3), 501–525 (2009)

Daems, K., De Pelsmacker, P., Moons, I.: The effect of ad integration and interactivity on young teenagers' memory, brand attitude and personal data sharing. Comput. Hum. Behav. **99**, 245–259 (2019)

Dahlen, M., Rosengren, S.: If advertising won't die, what will it be? Toward a working definition of advertising. J. Advert. **45**(3), 334–345 (2016)

Damiani, J.: Black Mirror: Bandersnatch could become Netflix's secret marketing weapon. The Verge (2019)

Daugherty, T., Djuric, V., Li, H., Leckenby, J.: Establishing a paradigm: a systematic analysis of interactive advertising research. J. Interact. Advert. **17**(1), 65–78 (2017)

Elnahla, N.: Black mirror: bandersnatch and how Netflix manipulates us, the new gods. Consumption Mark. Culture **23**(5), 1–6 (2019)

Frolova, S.: The Role of Advertising in Promoting a Product (2014)

Ghose, A., Singh, P.V., Todri, V.: Got annoyed? examining the advertising effectiveness and annoyance dynamics (2017)

Goldstein, D.G., Suri, S., McAfee, R.P., Ekstrand-Abueg, M., Diaz, F.: The economic and cognitive costs of annoying display advertisements. J. Mark. Res. **51**(6), 742–752 (2014)

Ha, L.: Online advertising research in advertising journals: a review. J. Curr. Issues Res Advert. **30**(1), 31–48 (2008)

Hu, Y., Shin, J., Tang, Z.: Pricing of online advertising: cost-per-click-through vs. cost-per-action. In: 2010 43rd Hawaii International Conference on System Sciences, pp. 1–9. IEEE (2010)

Jiang, M., McKay, B.A., Richards, J.I., Snyder, W.: Now you see me, but you don't know: consumer processing of native advertisements in online news sites. J. Interact. Advert. **17**(2), 92–108 (2017)

Keller, K.L.: Brand equity management in a multichannel, multimedia retail environment. J. Interact. Mark. **24**(2), 58–70 (2010)

Krishnamurthy, V., Aprem, A., Bhatt, S.: Multiple stopping time POMDPs: structural results & application in interactive advertising on social media. Automatica **95**, 385–398 (2018)

Kuester, S.: MKT 301: strategic marketing & marketing in specific industry contexts, p. 110. University of Mannheim (2012)

Langaro, D., Rita, P., de Fátima Salgueiro, M.: Do social networking sites contribute for building brands? evaluating the impact of users' participation on brand awareness and brand attitude. J. Mark. Commun. **24**(2), 1–23 (2015)

Li, H.: Special section introduction: artificial intelligence and advertising. J. Advert. **48**(4), 333–337 (2019)

Liu, Y., Shrum, L.J.: What is interactivity and is it always such a good thing? Implications of definition, person, and situation for the influence of interactivity on advertising effectiveness. J. Advert. **31**(4), 53–64 (2002)

Liu-Thompkins, Y.: A decade of online advertising research: what we learned and what we need to know. J. Advert. **48**(1), 1–13 (2019)

Madhavaram, S., Appan, R.: The potential implications of web-based marketing communications for consumers' implicit and explicit brand attitudes: a call for research. Psychol. Mark. **27**(2), 186–202 (2010)

Malthouse, E., Shankar, V.: A closer look into the future of interactive marketing. J. Interact. Mark. **23**(2), 105–107 (2009)

Michalski, E.: Interactive marketing management. Handel. Wewn. **5**(370), 289–297 (2017)

Ngowi, A.R.: The Effectiveness of Internet Advertising on Consumer Behaviour: The Case of Moshi Cooperative University Students. Doctoral dissertation, Mzumbe University (2015)

Nizam, N.Z.: Interactive online advertising: the effectiveness of marketing strategy towards customers purchase decision. Int. J. Hum. Technol. Interact. (IJHaTI) **2**(2), 9–16 (2018)

Park, C.S.: Interactive effects of advertising platform credibility and partisanship on advertising evaluation and recall. Asian J. Polit. Sci. 1–13 (2019)

Pavlou, P.A., Stewart, D.W.: Measuring the effects and effectiveness of interactive advertising: a research agenda. J. Interact. Advert. **1**(1), 61–77 (2000)

Ratchford, B.T.: Some directions for research in interactive marketing. J. Interact. Mark. **29**, v–vii (2015)

Roberts, M.S., Ko, H.: Global interactive advertising: defining what we mean and using what we have learned. J. Interact. Advert. **1**(2), 18–27 (2001)

Rodgers, S., Thorson, E.: Digital Advertising: Theory and Research. Routledge, New York (2017)

Schroll, R., Grohs, R.: Uncertainty in pre-release advertising. J. Advert. **48**(2), 167–180 (2019)

Söllner, J., Dost, F.: Exploring the selective use of ad blockers and testing banner appeals to reduce ad blocking. J. Advert. 1–11 (2019)

Speck, P.S., Elliott, M.T.: Predictors of advertising avoidance in print and broadcast media. J. Advert. **26**(3), 61–76 (1997)

Stewart, D.W., Pavlou, P.A.: From consumer response to active consumer: measuring the effectiveness of interactive media. J. Acad. Mark. Sci. **30**(4), 376–396 (2002)

Tudoran, A.A.: Why do internet consumers block ads? New evidence from consumer opinion mining and sentiment analysis. Internet Res. **29**(1), 144–166 (2019)

Wielki, J., Grabara, J.: The impact of ad-blocking on the sustainable development of the digital advertising ecosystem. Sustainability **10**(11), 4039 (2018)

Digital Business Models

Towards Different Enterprise Architecture Project Types

Aletta Klopper(⊠) ⓘ, Machdel Matthee ⓘ, and Alta van der Merwe ⓘ

Department of Informatics, University of Pretoria, Pretoria, South Africa
haklopper@gmail.com, {Machdel.Matthee,Alta.Vdm}@up.ac.za

Abstract. This research is in the enterprise architecture (EA) research field. EA is a developing discipline that in broad terms emphasizes all aspects of organizational design and development, including enabling information technology. However, there are various interpretations and understandings of EA, with little agreement on them. Therefore, organizations use EA in numerous ways to achieve different goals. These vary from purely information technology- (IT) related, internal business and IT-related to business environment-related goals. Enterprise architects also have different understandings of EA, which influence the way they perform EA work and consequently EA deliverables and achievement of EA project goals. In this paper a preliminary list of different EA project types is compiled through a hermeneutic literature review, aiming to establish a comprehensive list of EA project types. It is suggested that knowledge of different EA project types assist in the selection of suitable enterprise architects to achieve specific EA project goals.

Keywords: Enterprise architecture · Project · Project type

1 Introduction

Enterprise architecture (EA) can be defined as the components of which the enterprise is made up, how these components relate to each other and how they relate to the environment in which the enterprise operates, as well as the rules for their design and development over time [61]. However, architects do not have the same understanding of EA [20, 32] and it is not a "one-size-fits-all discipline" [64]. Perceptions vary between IT-focused, business-focused and a combination of business and IT, where business can also include the environment in which the organization operates [32].

Because of these different understandings, architects approach architecture work differently, which leads to misunderstandings and arguments about what EA processes to follow and which EA phases to perform [27]. The effect of these misunderstandings and different interpretations of EA and EA execution is that approaches to satisfy a requirement will be different, resulting in different EA designs [20]. Another implication is that architects who work together will differ on their roles and what they are responsible for. This may lead to conflict, which may complicate stakeholder engagement and EA project execution [50].

© Springer Nature Switzerland AG 2020
M. A. Bach Tobji et al. (Eds.): ICDEc 2020, LNBIP 395, pp. 141–154, 2020.
https://doi.org/10.1007/978-3-030-64642-4_12

There is more than one type of EA work. Korhonen and Poutanen [29] acknowledge that for each type of architecture work a unique approach, knowledge and skills set are required. They further mention that it is unrealistic to think that one person possesses all the required skills. This implies that enterprise architects with a combination of skills, knowledge and understanding of EA are required to complete EA projects of different types. It further emphasizes the importance of selecting enterprise architects with EA understanding, knowledge and skills relevant to the type of EA project to be undertaken. In this research we focus on the preliminary identification of different types of EA projects to be used in further research that will develop a method to assist the EA project manager with the selection of enterprise architects for EA project execution.

This paper is organized in five sections. In Sect. 1 background to the research study is provided. Section 2 gives an overview of EA, followed by Sect. 3 that describes the research methodology followed to determine EA project types. Section 4 contains the research result, i.e. the preliminary list of different EA project types that organizations can use to understand who to assign to the projects. Section 5 contains a discussion and the conclusion.

2 Enterprise Architecture

EA is valued by organizations as a discipline and practice to help them cope with continuous change [51] and to support decisions on organizational changes and relevant technology changes in support of business [35]. Responding to ongoing change is critical for organizational success. Therefore, it is important for organizations to take note of their enterprise architects' capabilities and views on EA, as this affects the way they practice as enterprise architects. Shaanika and Iyamu [57] state that the view on EA informs how EA is executed. This in turn has an impact on how well the organization responds to environmental changes that necessitate business and IT changes. The human component of an EA service capability is crucial for successful EA project execution [58].

According to Gartner [19], the time required to establish EA in an enterprise varies between 18 and 24 months and it takes an additional 12 to 24 months to improve and refine it. Apart from the time spent on EA, organizations also invest financially in EA. This is stressed by Bernard [14], who mentions that skilled enterprise architects, who come at a large cost, are required to develop architectures.

Development of EA artefacts is a labor-intensive, costly aspect of EA [45]. The people component of EA accounts for the larger part of the cost to establish, improve and maintain EA. It further emphasizes that enterprise architects with understanding of EA relevant to what the organization wants to achieve through execution of EA projects need to be identified and employed. Therefore, knowledge about the different EA project types is essential to ensure that the EA investment contributes to the success of the specific organization.

Effective EA implementation depends on the right type of person, with relevant skills, being employed to perform EA tasks [66]. By identifying and addressing human issues that influence the use and acceptance of EA as an organizational strategy, enterprises can prevent failure of their EA implementations [22].

Bakar and Hussien [12] identified five human-related factors that have an impact on EA execution. One of the human factors is skilled EA talent. Bakar and Hussien [12] do not provide a method to determine the required skills to perform EA work. However, Ylinen and Pekkola's [68] research focuses on identifying skills that enterprise architects themselves believe are crucial for performing EA work. They have found that the skills set to perform EA work is very broad and entails various separate tasks. In fact, 257 different skills were identified by the enterprise architects that participated in the study. This is due to different perceptions and experiences of EA. Ylinen and Pekkola [68] highlight the importance of selecting the right enterprise architect with the relevant skills for the specific EA project or phase of the project. Thus, organizations need to know what type of EA project is executed in order to select the most appropriate architects to work on it.

3 Research Methodology

3.1 Introduction

In this research the following three steps were followed to determine EA project types: (1) the definition and characteristics of a project were determined in order to identify EA project types, (2) a method was identified to determine project types from a literature review, and (3) a preliminary list of EA project types was derived by applying the method identified in Step 2. Figure 1 below illustrates the process followed.

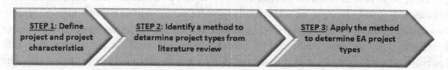

Fig. 1. Process followed to determine EA project types

3.2 Step 1: Define Project and Project Characteristics

The Project Management Institute describes a project as a "temporary endeavor undertaken to create a unique product, service, or result." Schwalbe [55] lists the unique purpose of a project as the first of six project characteristics. This unique purpose relates to the delivery of a specific product, service or result as per the Project Management Institute's definition of a project. The other project characteristics provided are that a project has a definite start and a definite end, projects can evolve over time and are performed in increments as more knowledge is gained, people and other resources are required to execute a project and a project has a key sponsor who normally dictates the project direction and provides funding. The last project feature mentioned is that there are unknown factors involved, such as time required to perform certain tasks and the availability of resources. The scope, time and cost of a project are referred to as the "triple constraint". Of these three constraints, the scope of a project defines what the project will deliver and what will be done to deliver it [55].

3.3 Step 2: Identify a Method to Determine Project Types from Literature Review

In a study in which method engineering processes are enhanced, Bucher et al. [18] distinguish between context and project type. They refer to the products that are developed through execution of projects as work systems, as their research is in the information systems subject field. It is stated that a work system includes all objects that are developed or transformed via a method. In the context of EA projects, EA project deliverables, delivered via a method, are similar to work systems.

A project type can be derived from the state of the original work system and the state of the target work system [18], which relates to the project scope as defined by [55]. By applying Bucher et al.'s [18] method for determining project types to the EA subject area, Aier et al. [3] give two examples of EA project types. The first example is the establishment of business processes and supporting information systems for a new business. The other example provided is the amalgamation of information systems that support business processes that are alike. Therefore, the way to determine project types focuses on what is delivered by or achieved through EA exercises, projects or initiatives, as the project type classification can be derived from it.

3.4 Step 3: Apply the Method to Determine EA Project Types

For Step 3 a literature review was performed with the focus on what is delivered through EA projects or initiatives. The literature review followed a hermeneutic approach, which implies that data collection and data analysis were performed simultaneously. Comprehension of literature was progressively enlightened through prior understanding of other literature, without being restricted by research protocols and formal approaches with specific rules in terms of articles that may or may not be included in the study [16]. The framework for hermeneutic literature review developed by Boell and Cecez-Kecmanovic [16] was applied. This framework prescribes two inter-linked cycles, i.e. "analysis and interpretation" as the broader cycle and "search and acquisition" as the internal cycle. Each cycle consists of specific phases that are performed iteratively, resulting in progressive better understanding of the literature. With each iteration through the hermeneutic circle, understanding of the topic is improved. Numerous iterations are done until the researcher reaches the point where a thorough recording of the literature is compiled, enabling analytical evaluation of the literature. The literature review starts with a primary topic, following the first iteration through the inner circle of searching, sorting, selecting, acquiring and reading. This is followed by the mapping and classifying, critical assessment and argument development phases of the first iteration through the outer circle. The inner circle continues when new literature is identified and search criteria are refined through reading, where after the next iteration through the inner circle starts, and then the next iteration through the outer circle.

Data collection through hermeneutic literature review is intertwined with hermeneutic data analysis techniques where the complete text is comprehended, where-after the researcher understands the complete text, interprets the parts that make up the whole, and then circles back to understanding the whole. With each cycle new insight is gained [36].

For this study more than 500 papers from three databases, i.e. Scopus, ScienceDirect and IEEE Xplore, were identified in several iterations through the hermeneutic circle. Initially, papers were identified through database searches using "enterprise architecture" and "enterprise architecture" + "project" in title, abstract and keywords as search criteria. Papers were scanned for relevance and an understanding of the EA project topic was gained, which triggered selection of more papers. Information gathered was compared to information previously gathered in order to identify EA project types. The relevance of papers was determined by applying the method to identify project types as described in paragraph 3.3. First the abstract was read; if it seemed as if the original work system vs target work system could be obtained, the paper was read. Again, it was determined if the original work system vs target work system could be obtained. If it could not be obtained, the paper was rejected.

4   EA Project Types

Twenty different EA project types, depicted in Table 1 below, were identified through the hermeneutic literature review.

Table 1. EA project types

No	EA project type	Source
1.	EA establishment project	[3, 46]
2.	Applying EA method to solve internal business problems	[65]
3.	Business-IT alignment	[4–7, 9, 11, 13, 15, 17, 21, 23, 25, 26, 30, 31, 33, 34, 39, 40, 43, 44, 48–50, 52, 56, 60, 62, 69]
4.	Business transformation	[1, 28, 37, 44]
5.	Digital transformation	[23]
6.	Improvement of organizational agility	[23, 31]
7.	Cost saving, including reduction in the cost of IT	[23, 31, 42, 50]
8.	Risk management	[23, 31]
9.	Enhancement of interoperability	[23, 42]
10.	Improvement in the results of strategic business programmes	[23]
11.	Business process optimisation	[23]
12.	Less complex IT systems	[23]
13.	Higher utilisation of IT systems	[23]
14.	Elimination of duplication of information systems	[3, 42, 50]
15.	Standardisation	[31, 54]
16.	Governance	[31, 54]
17.	Regulatory compliance	[31, 54]
18.	Corporate strategic planning	[6, 59]
19.	Organizational development	[14, 38, 47, 53]
20.	IT decision-making	[10, 62–64]

In the following paragraphs we reflect, using the literature review conducted, on how the EA project types (Table 1) were obtained from the sources. We provide the results from the data collection done during the literature review.

4.1 EA Project Type: EA Establishment Project

Pulkkinen and Kapraali [46] observe an EA establishment project during their study to develop a method to obtain information from information and communication technology (ICT) business users. The method is meant to be used for the development of EA. The EA initiation project is performed in a large organization that has the vision to implement EA. The method focuses on obtaining a business view first, not only by involving business managers but also by obtaining information regarding lower-level business operations and business processes. The original work system state is an organization without an EA function and the target work system state is an organization where EA is implemented. The EA project type identified establishes and deploys EA in an organization, similar to the first type of project mentioned by Aier et al. [3] when they applied the Bucher et al. [18] method to determine project types.

4.2 EA Project Type: Applying EA Method to Understand Internal Business Problems

Sometimes EA modelling and concepts are used for purposes other than to deliver EA. One such case is where Werewka and Spiechowicz [65] apply an EA approach to pinpoint problems with a specific process step in a Scrum agile development method for developing software. This process step is called the retrospective step. An EA approach is considered suitable for this exercise, as EA provides an all-inclusive view of an organization and describes the organization's future state and how to get to the future state. Further to this, EA models are known for providing different viewpoints relevant to different participants. Through different viewpoints stakeholders understand problems experienced in the agile software development approach and can determine the reasons why the agile software development method does not always achieve its goals [65]. In this case, EA concepts and modelling are applied to gain insight into a problem. The original work system state is a situation of experiencing a problem and the target work system state is a position of understanding the problem. This type of EA project is performed to understand problems experienced in specific business situations.

4.3 EA Project Type: Business-IT Alignment

A major challenge for enterprises to stay relevant and competitive is agility, which means that an organization must be agile to be able to react to constant and unanticipated change and to be able to integrate into the changing milieu in which it operates [13, 26]. EA improves organizational agility and by implication enhances organizational performance [24]. Enterprise agility is achieved when enterprise business and IT are kept aligned, despite changes in business and changes in technology. Business-IT alignment is achieved through enterprise engineering (EE) [23, 26]. As constant business-IT alignment is required, EE is not seen as a project, but as a continuous task [26]. EE and

EA are closely related disciplines. EE is described as the use of engineering theory to develop EA [41]. For the purpose of this study, EA with the purpose to align business and IT, and EE are seen as the same. Therefore, according to this argument, EA with the purpose to align business and IT is also not viewed as a project, as it is an ongoing effort with no specific start and end date. Keeping business and IT aligned through EA can be viewed as maintenance or sustainment of EA. The research of Agievich et al. [2] describes a method to create an EA baseline and to keep it current and in step with new IT solutions implemented in the organization. The motivation for using such a method is that one of the key success factors for a successful business, and to gain competitive advantage, is good administration and utilization of information by means of IT. This is addressed through EA. Agievich et al. [2] refer to the establishment of baseline EA and the maintenance of EA that follows to keep the baseline relevant. Establishment of baseline EA relates to the EA establishment and deployment project type derived from the work of Pulkkinen and Kapraali [46], and the sustainment of EA relates to the view that EA, with the purpose to align business and IT, is a continuous activity [26]. The EA project lifecycle has four main stages, namely "Initiate, Planning, Execute and Maintain" [8].

Each EA project will have these four phases. Where EA is seen as a continuous task [26] or reference is made to sustainment of EA [2], it relates to the maintenance phase in the EA project life cycle and is not considered a different type of EA project.

Lehong et al. [33] and Olsen [43] confirm that EA plays a large role in business-IT alignment. Business-IT alignment is achieved through EA when business, data, application and technology architecture is analyzed and understood. This refers to the business processes in support of organizational objectives, the information that will be created, replaced, updated and deleted by these business processes, applications that can manipulate the information, and the technology on which the applications execute [5, 39]. Antunes et al. [9] state that business-IT alignment can be achieved through EA models reflecting the business and IT components of the organization. Aldea et al. [6] identify EA development as a solution to align business and IT. They further stress the importance of this alignment, as it improves enterprise competitive advantage and organizational performance and ensures that strategic goals are supported and achieved through EA. A systematic literature review on research regarding business-IT alignment through EA points out that EA is perceived as a method to address business-IT alignment problems [69]. By utilizing the Zachman framework EA can be developed that ensures alignment of business objectives and IT [40]. Alaeddini et al. [4] mention that EA is a way to achieve business-IT alignment. They evaluate and measure the impact of executing EA on maturity of business-IT alignment. Their study confirms the positive effect of performing EA on business-IT alignment. Then again, EA maturity enhances business-IT alignment [17]. EA, specifically performed through a process-focused method, benefits business-IT alignment [34]. EA planning can be used to develop an architecture or ICT master plan that ensures integrated systems to support business goals [49, 67]. Business-IT alignment through EA is further recognized by several other authors, namely Alwadain et al. [7], Bakar et al. [11], Bhattacharya [15], Ernst [21], Hafsi and Assar [23], Hiekkanen et al. [25], Kotusev et al. [30], Lange and Mendling [31], Olsen and Trelsgård [44], Rouhani et al. [48], Schekkerman [52], Sessions [56], Tamm et al. [60] and Urbaczewski and Mrdalj [62].

Hence, one type of EA project is to achieve business and IT alignment. Although Hinkelmann et al. [26] argue that EE is not a project, their argument that EE (and thus also EA with the purpose to align business and IT) ensures business-IT alignment supports the business-IT alignment EA project type.

4.4 EA Project Type: Business Transformation

Nardello et al. [37] investigate how EA supports innovation. They view EA according to the three schools of thought on EA [32] and have found that the enterprise ecological adaption (EEA) school of thought is the only school of thought on EA that supports innovation. The reason for this is that the EEA school of thought includes the enterprise, the environment in which it operates and interaction between the enterprise and its environment.

Enterprises need to adapt to a changing environment; this transformation may be required due to internal or external incidents, such as emerging and disrupting technologies or new governance requirements. Enterprise transformation has an impact on more than one organizational unit, and it affects the enterprise relationships with one or more key stakeholders. Common understanding among the various role players in enterprise transformation is required. EA models contain the necessary information to establish common understanding between the different stakeholder groups. EA models contain information that spans stakeholder views, providing a holistic view of the organization [1]. EA is thus used for business transformation [44]; in fact, it is seen as a process that simplifies business transformation and integration [28]. EA is further viewed as a solution for enterprise integration into the changing environment in which it operates [13, 50]. The EA project type, business transformation, is derived from enterprise adaptation and integration into the dynamic environment and the need for agility and innovation.

4.5 EA Project Type: Digital Transformation

Related to the business transformation project type is the project type EA as a tool in support of digital transformation episodes. This type is derived from work done by Hafsi and Assar [23] to determine how EA can be used in support of digital transformation. They emphasize four areas where EA can contribute to digital transformation based on The Open Group Architecture Framework (TOGAF). These areas are the holistic view of all initiatives, the architecture vision that determines the scope of the project and ensures that business goals are addressed, the architecture repository and stakeholder management.

4.6 EA Project Type: Nine EA Project Types Derived from EA Benefits

Another product that was delivered as part of Hafsi and Assar's [23] study is a list of EA benefits obtained through a literature review. These benefits are improvement of organizational agility, business and IT alignment, a reduction in the cost of IT, better risk management and interoperability, improvement in the results of strategic business programs, business process optimization, less complex IT systems and higher utilization of these IT systems. In this study, each benefit is taken as a different EA project type, as each benefit represents what is achieved through an EA project. Of these EA project types business-IT alignment, organizational agility, cost saving and better risk management were included as EA goals when Lange and Mendling [31] extended Schöenherr's [54] list of EA goals. EA goals are realized through execution of EA projects and therefore each goal represents an EA project type. Ojo et al. [42] identify the realization of interoperability as a reason to perform EA in the public sector. Thus, it strengthens the concept of the EA project type, enabling interoperability.

The cost-saving EA project type is also derived from a study where reduction in business and IT cost through elimination of duplicate information systems and business processes is mentioned as a reason why EA is performed in the public sector [42]. A systematic mapping study of literature on various ways and reasons why EA is conducted confirms that EA can be performed to eliminate duplication in functionality and to enhance reuse of functionality in order to reduce IT cost [50]. Thus, elimination of duplication of information systems is another EA project type that is derived. Elimination of duplication of information systems relates to amalgamation of information systems that support business processes that are alike, as identified by Aier et al. [3] when they applied Bucher et al.'s [18] method for determining project types to the EA subject area.

4.7 EA Project Types: Three EA Project Types Derived from EA Goals

The list of EA goals compiled by Schöenherr [54] was extended by Lange and Mendling [31]. Apart from the EA goals already mentioned with the EA benefits identified by Hafsi and Assar [23], standardization, governance and regulatory compliance complete the extended list of EA goals [31]. Each of these goals signifies what is achieved through EA and may therefore represent an EA project type.

4.8 EA Project Type: Corporate Strategic Planning

Facilitation of strategic business planning is identified as an application of EA [6, 59], and therefore corporate strategic planning is identified as a potential EA project type.

4.9 EA Project Type: Organizational Development

Närman et al. [38] developed a framework that increases the application of EA to address organizational structure development. Enterprises use EA for organizational structuring in order to overcome business challenges [47]. EA is recognized as a discipline that aids in organizational structuring by delivering agile enterprise designs [14, 53]. Hence, another EA project type derived is organizational development.

4.10 EA Project Type: IT Decision-Making

One more application of EA is to assist in IT decision-making. EA improves the quality of IT decisions [64], as well as decisions on investments in IT [63]. EA projects are further used for IT decision-making by delivering blueprints for IT solutions [62] and providing a common understanding of the overall design of enterprise IT solutions [10].

5 Discussion and Conclusion

Table 1 reflects the 20 EA project types that were identified. The 20 different EA project types can further be categorized and similar types may be combined. One categorization that may be considered, as provided by Korhonen and Poutanen [29], is to categorize the project types as "technical, socio-technical or ecosystemic". An EA project classification framework can be developed using the identified EA project types. The EA classification framework, including skills required per project type, will benefit organizations in the selection of architects with skills and knowledge relevant to a specific EA project. Because of the many perceptions of EA [20], employing architects that match the requirements of the EA project type may enhance successful project execution.

Different perceptions of EA lead to different approaches to EA project execution, resulting in different EA designs and deliverables. Because of the immense cost and time invested in EA projects, and the large impact that EA has on organizations, it is not affordable to select architects to work on a project that do not match the requirements of the project. It is therefore necessary to be aware of the different EA project types that exist. Additional research must be done to determine characteristics of each EA project type, which will further aid in identifying enterprise architects with understanding of EA and the knowledge and skills relevant to the EA project type being executed.

References

1. Abraham, R., Aier, S., Winter, R.: Crossing the line: overcoming knowledge boundaries in enterprise transformation. Bus. Inform. Syst. Eng. **57**(1), 3–13 (2015). https://doi.org/10.1007/s12599-014-0361-1
2. Agievich, V., Taratukhin, V., Becker, J., Gimranov, R.: A new approach for collaborative enterprise architecture development. In: Proceedings - 2012 7th International Forum on Strategic Technology, IFOST 2012, pp. 1–5 (2013). https://doi.org/10.1109/ifost.2012.6357672
3. Aier, S., Riege, C., Winter, R.: Classification of enterprise architecture scenarios - an exploratory analysis. Enterp. Model. Inform. Syst. Archit. **3**(1), 14–23 (2008)
4. Alaeddini, M., Asgari, H., Gharibi, A., Rad, M.R.: Leveraging business-IT alignment through enterprise architecture—an empirical study to estimate the extents. Inf. Technol. Manage. **18**(1), 55–82 (2016). https://doi.org/10.1007/s10799-016-0256-6
5. Alaeddini, M., Salekfard, S.: Investigating the role of an enterprise architecture project in the business-IT alignment in Iran. Inform. Syst. Front. **15**(1), 67–88 (2013). https://doi.org/10.1007/s10796-011-9332-y
6. Aldea, A., Iacob, M.E., Quartel, D., Franken, H.: Strategic planning and enterprise architecture. In: Proceedings of the First International Conference on Enterprise Systems, ES 2013, pp. 1–8. IEEE (2013). https://doi.org/10.1109/es.2013.6690089
7. Alwadain, A., Rosemann, M., Fielt, E., Korthaus, A.: Enterprise architecture and the integration of service oriented architecture. In: PACIS 2011 - 15th Pacific Asia Conference on Information Systems: Quality Research in Pacific (2011)

8. Anajafi, F., Nassiri, R., Shabgahi, G.L.: Developing effective project management for enterprise architecture projects. In: ICSTE 2010 - 2010 Second International Conference on Software Technology and Engineering, Proceedings. pp. V1-388-V1-393. IEEE (2010). https://doi.org/10.1109/icste.2010.5608867

9. Antunes, G., Bakhshandeh, M., Mayer, R., Borbinha, J., Caetano, A.: Using ontologies for enterprise architecture analysis. In: 2013 17th IEEE International Enterprise Distributed Object Computing Conference Workshops, pp. 361–368 (2013). https://doi.org/10.1109/edocw.2013.47

10. Armour, F.J., Kaisler, S.H., Liu, S.Y.: A Big-picture Look at Enterprise, pp. 35–41. IT Professional, Jan/Feb (1999)

11. Bakar, N.A.A., Harihodin, S., Kama, N.: Enterprise architecture implementation model: measurement from experts and practitioner perspectives. In: 4th IEEE International Colloquium on Information Science and Technology. pp. 1–6 (2016)

12. Bakar, N.A.A., Hussien, S.S.: Association of people factors with successful enterprise architecture implementation. Int. J. Eng. Technol. 7(4.31), 52–57 (2018)

13. Banaeianjahromi, N., Smolander, K.: What do we know about the role of enterprise architecture in enterprise integration? A systematic mapping study. J. Enterp. Inform. Manage. 29(1), 140–164 (2016). https://doi.org/10.1108/jeim-12-2014-0114

14. Bernard, S.A.: An Introduction to Enterprise Architecture. AuthorHouse (2012)

15. Bhattacharya, P.: Modelling strategic alignment of business and IT through enterprise architecture: augmenting archimate with BMM. Procedia Comput. Sci. 121, 80–88 (2017). https://doi.org/10.1016/j.procs.2017.11.012

16. Boell, S.K., Cecez-Kecmanovic, D.: A hermeneutic approach for conducting literature reviews and literature searches. Commun. Assoc. Inform. Syst. 34(12), 257–286. https://doi.org/10.17705/1cais.03412

17. Bradley, R.V., Pratt, R.M.E., Byrd, T.A., Outlay, C.N., Wynn, D.E.: Enterprise architecture, IT effectiveness and the mediating role of IT alignment in US hospitals. Inform. Syst. J. 22(2), 97–127 (2012). https://doi.org/10.1111/j.1365-2575.2011.00379.x

18. Bucher, T., Klesse, M., Kurpjuweit, S., Winter, R.: Situational method engineering. In: Ralyté, J., Brinkkemper, S., Henderson-Sellers, B. (eds.) Situational Method Engineering: Fundamentals and Experiences. ITIFIP, vol. 244, pp. 33–48. Springer, Boston, MA (2007). https://doi.org/10.1007/978-0-387-73947-2_5

19. Burke, B., Blosch, M.: ITScore Overview for enterprise architecture. Gartner Anal. 1–6 (2015)

20. Du Preez, J.A.: Understanding the architect in enterprise architecture: The Daedalus instrument for architects. Doctoral thesis. Pretoria: University of Pretoria (2016). https://repository.up.ac.za/handle/2263/57172. Accessed 04 Mar 2020

21. Ernst, A.M.: Enterprise architecture management patterns. In: 15th Conference on Pattern Languages of Program (PLoP), pp. 1–20 (2008)

22. Gilliland, S., Kotzé, P., Van Der Merwe, A.: Work level related human factors for enterprise architecture as organisational strategy. In: 2015 International Conference on Enterprise Systems (ES), pp. 43–54. IEEE (2015). https://doi.org/10.1109/es.2015.12

23. Hafsi, M., Assar, S.: What enterprise architecture can bring for digital transformation? An exploratory study. In: IEEE 18th Conference on Business Informatics. pp. 83–89 (2016). https://doi.org/10.1109/cbi.2016.55

24. Hazen, B.T., Bradley, R.V., Bell, J.E., In, J., Byrd, T.A.: Enterprise architecture: a competence-based approach to achieving agility and firm performance. Int. J. Prod. Econ. 193(July), 566–577 (2017). https://doi.org/10.1016/j.ijpe.2017.08.022

25. Hiekkanen, K., Korhonen, J.J., Collin, J., Patricio, E., Helenius, M., Mykkanen, J.: Architects' perceptions on EA use - An empirical study. In: Proceedings - 2013 IEEE International Conference on Business Informatics, IEEE CBI 2013. pp. 292–297 (2013). https://doi.org/10.1109/cbi.2013.48

26. Hinkelmann, K., Gerber, A., Karagiannis, D., Thoenssen, B., Van Der Merwe, A., Woitsch, R.: A new paradigm for the continuous alignment of business and IT: combining enterprise architecture modelling and enterprise ontology. Comput. Ind. **79**, 77–86 (2016). https://doi.org/10.1016/j.compind.2015.07.009

27. Iyamu, T.: Institutionalisation of enterprise architecture: The actor-network perspective. In: Social and Professional Applications of Actor-Network Theory for Technology Development, pp. 144–155. IGI Global (2013)

28. Kaddoumi, T., Watfa, M.: A proposed agile enterprise architecture framework. In: The Sixth International Conference on Innovative Computing Technology, pp. 52–57 (2016)

29. Korhonen, J.J., Poutanen, J.: Tripartite approach to enterprise architecture. J. Enterp. Arch. **9**(1), 28–38 (2013)

30. Kotusev, S., Singh, M., Storey, I.: Investigating the usage of enterprise architecture artifacts. In: 23rd European Conference on Information Systems (ECIS), pp. 1–12 (2015)

31. Lange, M., Mendling, J.: An experts' perspective on enterprise architecture goals, framework adoption and benefit assessment. In: 2011 IEEE 15th International Enterprise Distributed Object Computing Conference Workshops, pp. 304–313. IEEE (2011). https://doi.org/10.1109/edocw.2011.41

32. Lapalme, J.: Three schools of thought on enterprise architecture. IT Prof. **14**(6), 37–43 (2012). https://doi.org/10.1109/mitp.2011.109

33. Lehong, S.M., Dube, E., Angelopoulos, G.: An investigation into the perceptions of business stakeholders on the benefits of enterprise architecture: the case of Telkom SA. S. Afr. J. Bus. Manage. **44**(2), 45–56 (2013)

34. Malta, P., Sousa, R.D.: Process oriented approaches in enterprise architecture for business-IT alignment. Procedia Comput. Sci. **100**, 888–893 (2016). https://doi.org/10.1016/j.procs.2016.09.239

35. Microsoft (2015). https://msdn.microsoft.com/en-us/library/ms978007.aspx. Accessed 03 Oct 2015

36. Myers, M.D.: Qualitative Research in Business and Management. Sage, London (2013)

37. Nardello, M., Lapalme, J., Toppenberg, G., Gøtze, J.: How does enterprise architecture support innovation? In: 2015 Third International Conference on Enterprise Systems, pp. 192–199 (2015). https://doi.org/10.1109/es.2015.26

38. Närman, P., Johnson, P., Gingnell, L.: Using enterprise architecture to analyse how organisational structure impact motivation and learning. Enterp. Inform. Syst. (June 2015), 1–40 (2014). https://doi.org/10.1080/17517575.2014.986211

39. Niemi, E., Pekkola, S.: Using enterprise architecture artefacts in an organisation. Enterp. Inform. Syst. **11**(3), 313–338 (2017). https://doi.org/10.1080/17517575.2015.1048831

40. Nogueira, J.M., Romero, D., Espadas, J., Molina, A.: Leveraging the Zachman framework implementation using action–research methodology – a case study: aligning the enterprise architecture and the business goals. Enterp. Inform. Syst. **7**(1), 100–132 (2013). https://doi.org/10.1080/17517575.2012.678387

41. Nurcan, S., Schmidt, R.: Service oriented enterprise architecture for enterprise engineering. In: 2013 17th IEEE International Enterprise Distributed Object Computing Conference Workshops, pp. 91–93. IEEE (2013). https://doi.org/10.1109/edocw.2013.49

42. Ojo, A., Janowski, T., Estevez, E.: Improving government enterprise architecture practice - Maturity factor analysis. In: Proceedings of the Annual Hawaii International Conference on System Sciences, pp. 4260–4269. IEEE (2012). https://doi.org/10.1109/hicss.2012.14

43. Olsen, D.H.: Enterprise architecture management challenges in the Norwegian health sector. Procedia Comput. Sci. **121**, 637–645 (2017). https://doi.org/10.1016/j.procs.2017.11.084

44. Olsen, D.H., Trelsgård, K.: Enterprise Architecture adoption challenges: an exploratory case study of the Norwegian higher education sector. Procedia Comput. Sci. **100**(1877), 804–811 (2016). https://doi.org/10.1016/j.procs.2016.09.228

45. Perez-Castillo, R., Ruiz-Ggonzalez, F., Genero, M., Piattini, M.: A systematic mapping study on enterprise architecture mining. Enterp. Inform. Syst. **13**(5), 675–718 (2019). https://doi.org/10.1080/17517575.2019.1590859
46. Pulkkinen, M., Kapraali, L.: Collaborative EA information elicitation method: the IEM for business architecture. In: Proceedings - 17th IEEE Conference on Business Informatics, CBI 2015, pp. 64–71 IEEE (2015). https://doi.org/10.1109/cbi.2015.33
47. Rajabi, Z., Minaei, B., Seyyedi, M.A.: Enterprise architecture development based on enterprise ontology. J. Theor. Appl. Electron. Commer. Res. **8**(2), 85–95 (2013). https://doi.org/10.4067/s0718-18762013000200007
48. Rouhani, B.D., Mahrin, M.N., Nikpay, F., Ahmad, R.N.: A systematic literature review on enterprise architecture implementation methodologies. Inf. Softw. Technol. **62**, 1–20 (2015). https://doi.org/10.1016/j.infsof.2015.01.012
49. Ruldeviyani, Y., Wisnuwardhani, E., Sucahyo, Y.G.: Designing enterprise architecture: case study of the ministry of energy and mineral resources. J. Eng. Appl. Sci. **12**(8), 2185–2188 (2017)
50. Saint-Louis, P., Lapalme, J.: An exploration of the many ways to approach the discipline of enterprise architecture. Int. J. Eng. Bus. Manage. **10**, 1–26 (2018). https://doi.org/10.1177/1847979018807383
51. Saint-Louis, P., Morency, M.C., Lapalme, J.: Defining enterprise architecture: a systematic literature review. In: 2017 IEEE 21st International Enterprise Distributed Object Computing Workshop, pp. 41–49 (2017). https://doi.org/10.1109/edoc.2017.16
52. Schekkerman, J.: How to Survive in the Jungle of Enterprise Architecture Frameworks: Creating or Choosing an Enterprise Architecture Framework. Trafford, Victoria B.C. (2004)
53. Schelp, J., Stutz, M.: A balanced scorecards approach to measure the value of enterprise architecture. In: Trends in Enterprise Architecture Research Workshop, pp. 5–12 (2007)
54. Schöenherr, M.: Towards common terminology in the discipline of Enterprise Architecture. Lect. Notes Comput. Sci. **5472**, 400–413 (2009)
55. Schwalbe, K.: Information technology project management. 7th edn. Course Technology, Cengage Learning (2014)
56. Sessions, R.: A comparison of the top four enterprise architecture methodologies, pp. 1–31. Msdn (2007)
57. Shaanika, I., Iyamu, T.: Developing enterprise architecture skills: a developing country perspective. In: International Federation for Information Processing (IFIP), pp. 52–61 (2014)
58. Shanks, G., Gloet, M., Someh, I.A., Frampton, K., Tamm, T.: Achieving benefits with enterprise architecture. J. Strateg. Inform. Syst. **27**(March), 139–156 (2018). https://doi.org/10.1016/j.jsis.2018.03.001
59. Simon, D., Fischbach, K., Schoder, D.: Enterprise architecture management and its role in corporate strategic management. IseB **12**(1), 5–42 (2013). https://doi.org/10.1007/s10257-013-0213-4
60. Tamm, T., Seddon, P.B., Shanks, G., Reynolds, P.: How does enterprise architecture add value to organisations? Commun. Assoc. Inform. Syst. **28**(10), 141–168 (2011)
61. The Open Group. The TOGAF® Standard, Version 9.2 (2018)
62. Urbaczewski, L., Mrdalj, S.: A comparison of enterprise architecture frameworks. Issues Inform. Syst. VII **2**, 18–23 (2006)
63. Van den Berg, M., Slot, R., Van Steenbergen, M., Faasse, P., Van Vliet, H.: How enterprise architecture improves the quality of IT investment decisions. J. Syst. Softw. **152**, 134–150 (2019). https://doi.org/10.1016/j.jss.2019.02.053
64. Van Den Berg, M., Van Vliet, H.: The decision-making context influences the role of the enterprise architect. In: IEEE 20th International Enterprise Distributed Object Computing Workshop (EDOCW), pp. 1–8 (2016)

65. Werewka, J., Spiechowicz, A.: Enterprise architecture approach to SCRUM processes, sprint retrospective example. In: Proceedings of the 2017 Federated Conference on Computer Science and Information Systems. pp. 1221–1228 (2017). https://doi.org/10.15439/2017f96

66. Wißotzki, M., Timm, F., Stelzer, P.: Current state of governance roles in enterprise architecture management frameworks. In: Johansson, B., Møller, C., Chaudhuri, A., Sudzina, F. (eds.) BIR 2017. LNBIP, vol. 295, pp. 3–15. Springer, Cham (2017). https://doi.org/10.1007/978-3-319-64930-6_1

67. Wikusna, W.: Enterprise architecture model for vocational high school. Int. J. Appl. Inform. Technol. **02**(01), 22–28 (2018). https://doi.org/10.25124/ijait.v2i01.925

68. Ylinen, M., Pekkola, S.: Looking for a five-legged sheep: Identifying enterprise architects' skills and competencies. In: 19th Annual International Conference on Digital Government Research: Governance in the Data Age, pp. 1–8 (2018). https://doi.org/10.1145/3209281.320 9353

69. Zhang, M., Chen, H., Luo, A.: A systematic review of business-IT alignment research with enterprise architecture. IEEE Access **6**, 18933–18944 (2018). https://doi.org/10.1109/access.2018.2819185

Digital Business Model Innovation in SMEs - Case Studies with DIH Support from Brandenburg (Germany)

Marc Gebauer[1]($\boxtimes$), Cyrine Tangour[2,3], and Diana Zeitschel[4]

[1] Brandenburg University of Technology, Siemens-Halske-Ring 14, 03046 Cottbus, Germany
marc.gebauer@b-tu.de
[2] Fraunhofer Center for International Management and Knowledge Economy - IMW,
Leipzig, Germany
[3] Faculty of Economics and Management Science,
Leipzig University, Leipzig, Germany
[4] Innovation Center Modern Industry Brandenburg - IMI Brandenburg,
Siemens-Halske-Ring 14, 03046 Cottbus, Germany

Abstract. Digital technologies provide opportunities for diverse companies to innovate their business models in different ways. The process of business model innovation triggered by digitization is especially challenging for small and medium-sized enterprises (SMEs). In addition, literature on business model innovation, following the digitization of SMEs business process, remains limited. Thus, the article uses an exploratory case study approach to analyze two different business model innovation processes triggered by digital technologies. Additionally, the SMEs benefited from the support of a regional digital innovation hub for digitizing their processes. In this paper, the authors discuss two business model innovation paths, growth-focused and efficiency-focused paths.

Keywords: Digital business model · SME · Business model innovation · Brandenburg · Digital innovation hub

1 Introduction

Digital technologies are pushing innovation of companies from different geographic regions and industries [1]. Consequently, the integration of digital technologies in firms' processes hold the potential to change every element of a business model and thus, for enabling business model innovation (BMI) [2]. A BMI that is difficult to imitate due to its complex articulation can create a strong competitive advantage for the firm. In fact, in the recent publications, many definitions for business model and BMI emerged. BMI can be viewed as process or an outcome. When BMI is understood as an outcome, scholars tend to focus on more descriptive aspects of novel business models [3]. This research stream has a plethora of tools to capture empirically elements of companies business models, as by the accepted generic model of the Business Model Canvas [4].

© Springer Nature Switzerland AG 2020
M. A. Bach Tobji et al. (Eds.): ICDEc 2020, LNBIP 395, pp. 155–165, 2020.
https://doi.org/10.1007/978-3-030-64642-4_13

Other scholars are more inclined to adopt a dynamic approach to explore various aspects that drive or hinder the process of BMI [15].

However, the literature lacks empirical analysis and also tools for the so-called dynamic perspective or the process of BMI [3, 30]. In addition, studies that have an empirical focus, often center their attention rather on big companies than SMEs. More specifically the analysis of BMI processes in SMEs and longitudinal case studies are scarce. Especially, studies describing the adaption of digital technologies by SMEs can be further developed [5]. Actually, empirical research on SMEs is very important due to their significance for the economy of different regions [6–9]. The role of SMEs in the economic growth of the German state of Brandenburg is significant for many reasons. More than 90% of companies in Brandenburg belong to the category of SMEs [10]. Those companies are in need of innovation, especially based on digitization to stay competitive [11]. Consequently, this exploratory article focusses on the process of digital-based BMI in Brandenburg's SMEs. In addition, the authors describe the supportive role of digital innovation hub throughout the iterative steps of the SME's innovation paths, the problems they encountered and the nature of the assistance the digital innovation hub provided them with.

Therefor the next section provides an overview of former work about the implementation of digital technologies in SMEs of Brandenburg and scholarly discussions in the field of digital BMI. Afterwards, in Sect. 3 the authors describe the case study design. Section 4 displays the case studies results. Section 5 contains the discussion and Sect. 6 the conclusion.

2   Literature Review

2.1   Digital Business Model Innovation in SMEs

A business model has different meanings [3]. According to [4] and [12] business models are the logic of a company's value creation, value capture and value delivery which can be described by a system of business model elements. To capture these elements, many tools in the form of generic frameworks provide a delimited structure for capturing and analyzing business models [13]. For instance, the Business Model Canvas which belongs to the best-established generic business model frameworks (see e.g. [14]), has been often used in academia and the practice to devise business model innovation. In fact, scholars perceive BMI from at least two main perspectives: as result or as a process [3]. In this article, the authors adopt the process-based view of BMI. Thus, BMI is "...*a process that deliberately changes the core elements of a firm and its business logic...*" [23]. Several authors already defined sub-processes which emphasizes the importance of the process oriented perspective [16]. Others differentiate between evolutionary and revolutionary BMI [3, 17]. This article's focus lies on evolutionary BMI, which is rather an incremental innovation process evolving in minor steps over a relatively stretched period. Research on business models accelerated with the internet boom. Indeed, digital technologies provide opportunities to innovate business models to achieve competitive advantages. Companies are enabled e.g. to improve customers experiences, increase the value created or offer new value propositions to customers. Relying on the business model definition of this article the process of digital BMI is achieved when digital technologies change value creation, value capturing or value delivering in a business model. The process of BMI admits severe challenges for all companies alike, since conflicts may raise between the existing and the new business model elements. More particularly, SMEs face other specific challenges during the process of BMI because of specific inherent characteristics [18]. Unlike large companies SMEs have: *"personalised management, with little devolution of authority; severe resource limitations in terms of management, strategic capabilities, and finance; high specialization applied to a narrow range of products/services; reliance on a small number of customers; niche market orientation; flat, flexible organizational structures; high innovation potential; reactive to environment changes and legislative reforms; informal and unstructured strategy design"* [19]. For SMEs, characteristic barriers to innovation are problems in networking capacities as well as strategic capabilities, vision and resources for growth [20–22]. Fortunately, institutional support for instance in the form of the company's network can mitigate several of these BMI related challenges.

Since the authors put their attention on the process of changing business model elements resulting in evolutionary BMI, innovation paths for business models and their elements are a promising approach [23]. BMI paths describe different sequences of changed business model elements. Heikkilä et al. argue in their SME-focused exploratory article that the paths correlate with strategic goals. The approach of BMI paths is seen to be specially suitable for studying SMEs because of *"...often sequential, non-linear and iterative steps..."* which they perform to innovate their business models [24]. Growth, profitability and starting a new company are distinguished as strategic goals for initiating a BMI. Since the authors work with existing businesses, the only the strategic goals of growth and profitability are relevant. So-called growth seekers are said to start their BMI by changing business model elements from the right-hand sight of the business model canvas. Profitability oriented companies, in contrast, tend to start at the left-hand sight (Fig. 1). Heikkilä et al. found that, on the one hand, companies focusing on growth start with attracting new customers (channels) and go on with means to get to know their customer better (customer segments), the improvement of the offering (value proposition) and afterwards, adapt the elements of value creation like key partners. On the other hand, cases of profitability driven SMEs try to improve their efficiency by changing key processes, key activities and cost structure first [4, 24].

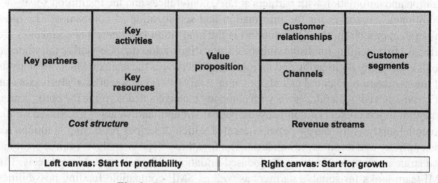

Fig. 1. Business model canvas based on [4]

2.2 Implementation of Digital Technologies in SMEs in Brandenburg

The economy in Germany as well as in its state Brandenburg is characterized by an entrepreneurial landscape largely dominated by SMEs. According to the European Commission 99.5% of companies with 54.4% of value added and 63,3% of the persons employed belong to the category of SMEs [25].

When digitizing analogue business processes which the business model elements are characterized by [26] the SMEs generally face problems like the lack of knowledge about the specific technologies required, education and skills of managers and employees as well as funding this innovation [27, 28]. Based on a survey of 68 German production SMEs, Müller et al. found specific concerns of SMEs regarding *"...substantial orga-nizational effort..."*, *"...high investments in machine parks and IT infrastructure [and] IT personnel and technical trainings..."*, *"...data security..."*, *"...small batch size..."* unsuitable for cost efficient automation in general, *"...varying automation degrees and lifecycle stages of machinery..."*, *"...putting information into commercial use..."* and *"...highly individual customer demands..."* [5]. SME's Brandenburg face similar problems [11]. The process of digitization in SMEs is characterized as incremental [29] or in case of the Industry 4.0 implementation as an *"...evolutionary process..."* [5].

An SME's specific problems and limitations result in the need for partnerships with other companies as well as academic and governmental institutions for choosing appropriate digital technologies to be implemented into their business model (see e.g. [18]). Thus, they need to be anchored in an innovation ecosystem as defined be the triple helix model, which is consisting of three main pillars: industry, academia and government [31]. In practice the triple helix model can be implemented in various ways. For instance, the "Innovation Centre of Modern Industry Brandenburg" (IMI Brandenburg) is a transfer project financed by the European Regional Development Funds that supports SMEs in Brandenburg through digitization and automation projects (www.efre.brandenburg.de), [32]. The IMI Brandenburg is embedded in a network of universities, solution partners, chambers and other DIHs across the European Union and provides companies with a model factory, a knowledge forum and an innovation lab. In the model factory companies are shown e.g. robot-assisted automation solutions and augmented reality technologies. The knowledge forum is set up to share and jointly develop knowledge concerning digitization and automation with partners at fairs, conferences and the institution's website. Additionally, it serves for the information and sensitization of companies. The most in-depth approach for the SME-support is the innovation lab. Interested companies are offered a first meeting for discussing the current digitization status as well as the strategic goals concerning digitization and automation. Afterwards the team conducts an analysis of the digitization potential called "potential analysis". This potential analysis consists of an on-site visit with interviews with company managers to determine the optimization potential in the context of company-individual strength and weaknesses. Based on the potential analysis a strategy paper is created which describes promising solutions for the company and its strategic situation. The first three steps of analysis (goals, potential and strategic analyses) support the focused initiation of digital innovation projects. The IMI-team looks for suitable partners as well as SME-compatible funding possibilities (www.imi4bb.de) for the implementation of the digital innovation project. Both cases study companies elaborated in this paper benefited from the support of IMI-Brandenburg to initialise their digitization project.

Kampe et al. state that Brandenburg is generally at the same level of digitization as other states in Germany. Nonetheless, especially micro and small companies show potential in digitizing their processes and business models. The level of digitization differs between diverse industries, too. Handcraft businesses recently show the strongest efforts in digitization. Also, in knowledge intense services the influence of digital technologies is already strong. In contrast to production processes, back-office processes show a higher degree of digitization in the SMEs [11]. According to a survey most companies aim at realizing digital potentials in logistics and supply chain management (60%), factory layout planning (50%), production planning and control systems (50%), automation (50%) and enterprise resource planning (40%) [18]. A similar study found enterprise resource planning, automation and factory planning to be the specific firm's processes where problems were named most often [33]. Digitization has positive effects on productivity, development of new products and services and profit increase [11].

One example of digitization degrees for manufacturing SMEs is given by Müller et al. with a focus on Industry 4.0. They consider four stages. Companies at the first stage do not consider digitization efforts and rely on their known processes. At the second stage, the SMEs have not decided for a technology but consider changes based on those technologies. At the third stage companies have already implemented technologies e.g. in production processes or for channels to customers. The fourth and final stage is companies aiming for the lead in their industry by full-scale adoption of available technologies for the digitization of their business model [5].

3   Methods and Design

3.1   Case Study Design

Case studies are suitable for analyzing complex objectives which cannot be approached with quantitative methods on a standardized basis [34]. Questions like "*how*" and "*why*" are supposed to be answered [35].

The article focusses on the question of how processes of digital-based BMI are happening in SMEs in Brandenburg. This work focusses on BMI paths, problems and the support that SMEs received from the digital innovation hub IMI Brandenburg. Thus, a case study approach fits the research. Case study research generally consists of the following steps: plan, design, prepare, collect, analyze and share [36]. The planning of the case study approach is to be seen in the context of the goals of the IMI Brandenburg in helping regional SMEs conducting their digitization approaches. Therefor digitization processes including problems and priorities of the SMEs in Brandenburg need to be understood. The design of the case study includes the analysis of two SMEs including the company goals, problems, support by IMI Brandenburg and the resulting business model innovation paths. To improve the case study quality the authors apply triangulation and rely on primary and secondary data sources. Primary data has been collected while working with the companies. Different documents like protocols, strategy papers and potential analysis which are developed by IMI Brandenburg have been used as forms of record of the semi structured interviews with the companies in different stages of their innovation processes. The company web sites as well as newspaper articles provide secondary data. The preparation as well as the collection were included in the processes of IMI Brandenburg's work. This article serves for the analysis as well as to share the results.

4 Case Study Analysis

4.1 Case A: Company A - Growth Orientation

Company A produces a great variety of frames including customer specific versions for pictures, TVs and other elements of daily life. Different materials increase the number of possibilities for customers to find or create their desired frame. It therefore reduces the batch size, too. From order to delivery it takes between three and five days. The company was bought about ten years ago by the current owner with no employees. Currently about 25 employees produce daily on average 100 frames and up to a maximum of 300 frames. It grows about 30% per year and has reached a yearly turnover of 1.5 Mio. € in 2016. Figure 2 shows the BMI path of this company based on the gradual introduction of digital technologies.

Case A - Growth

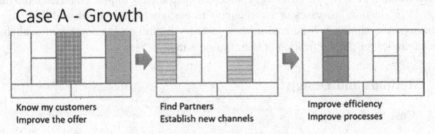

Know my customers   Find Partners   Improve efficiency
Improve the offer   Establish new channels   Improve processes

Fig. 2. Business model innovation path of case A

Starting in 2011 with one employee, the first goal was to find new customers and create a steady turnover. Thus, an online shop as a new delivery channel was implemented. In 2013 the shop was launched and a year later an enterprise resource planning (ERP) system for a better organization of key activities was introduced. It included data for 70.000 products and was supposed to be connected with the production planning system as well as with the merchandise management system. However, production processes were not adapted to deliver a great variety of frames. Flexible handcraft-based processes showed not to be sufficient anymore to respond to the demand generated by the online shops. This is why the company initiated an analysis project with a professorship for factory planning (from the IMI Brandenburg network) with the aim to determine inefficiencies of current inhouse logistics and to create a new factory layout. A virtual reality lens helped showing the results of the process analysis to the employees and was used a support for discussing the improvement ideas. A special regional funding for technology transfer was used. The next step of the company's innovation process was to connect the ERP system with the own online shop as well as other platforms the company uses to sell its products. The company contacted the IMI Brandenburg for support in identifying suitable digital solutions to connect their ERP system to its various delivery channels. Additionally, a check-up of the whole value chain was conducted by a partner from the IMI Brandenburg's network, a strategy paper developed and ideas for new projects created with IMI Brandenburg. The company pointed out that working with IMI Brandenburg is especially useful to find the right partners from university. Strategic goals also included growth of the B2B and the B2C business via the online channel, improving production planning, service and order tracking relying on production data collection and possibly the automation of a workstation at this point of time. The expansion of international customer segments partly funded and supported by a regional chamber in a project, which again extended the online channels.

4.2 Case B: Company B - Profitability Orientation

Company B is a specialist for complex and precise turn and milling parts. B2B customers include companies from aviation, medicine, science and telecommunication. The family run business has a turnover of about 4 Mio. € per year and employs about 30 people. A good proportion of the 5,000 parts the company produces every year are single-unit products. The company regards efficiency and digitization as a basis for growth and prioritizes innovation processes accordingly.

Since the company was founded in 2010, digitalization has been seen as a suitable means of realizing profitable and efficient product processes. With the initially 10 employees, the first step was to build up a modern production facility with programmable computerized numerical control (CNC) and connective machinery. A computer-aided design and manufacturing (CAD/CAM) solution was introduced to design the products and program the machining centers. An ERP system was introduced in 2015, which enabled the acquisition of more than 1,000 materials and assigned orders to the corresponding employees and machines. This made production planning and order tracking possible. The accounting office was equipped with a document management software including a link to the tax consultants. The cooperation with IMI Brandenburg resulted in a strategy paper for transparency considering tool management and production planning systems. Suggestions for integrating tool management into production and recording order status at terminals were developed, and suitable partners for implementation were recommended.

Every single innovation process in the company's value chain is implemented complementarily to the others. Figure 3 gives an overview. After successfully establishing digital internal processes, the company was able to offer new products to customers, e.g. the manufacture of high-precision components for space travel and research. New services relying on online channels for inquiries and offers were developed as well. The company has sought technology partners to implement these new processes and a suitable project manager has been found by IMI Brandenburg.

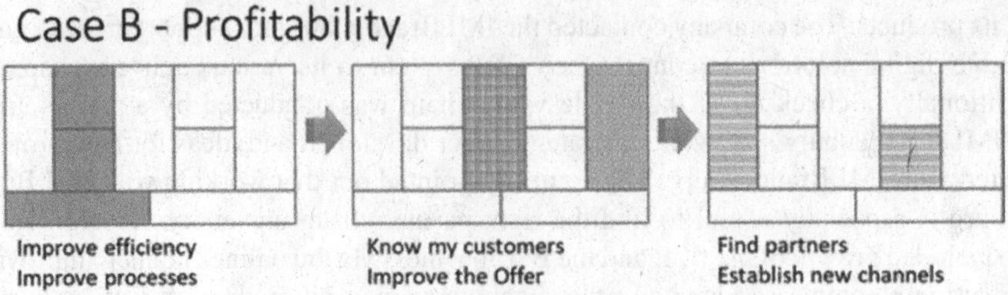

Case B - Profitability

Improve efficiency   Know my customers   Find partners
Improve processes   Improve the Offer   Establish new channels

Fig. 3. Business model innovation path of case B

5 Discussion

Both cases show that digital BMI in the SMEs should be regarded as an iterative process of several steps or an evolutionary BMI. Both companies implement digital technologies in several business model elements aiming at a completely digitized value chain over a period of several years.

Considering the SME specific problems for the process of digital business model innovation the cases in this paper by having implemented digital technologies in production processes and customer channels. They have a vision of an ongoing integration of digital technologies in their business model elements to become a market leader. Thus, they are aiming for the 4th and last stage of the digitization model by Müller et al. [5]. SME specific problems indicated by the cases include building a network and creating transparency about digital solutions for the companies' strategic goals. The support of the digital innovation hub IMI Brandenburg comprised prioritizing strategic goals, market research on digital solutions and finding partners according to the companies' needs.

The growth-oriented company prioritizes the delivery channels but focusses also at efficiency of production processes via digitization. The company B, which is more profitability-oriented starts via efficiency projects altering digitalization in key activities and key resources. Growth is also achieved in the second case. Thus, the case studies support the results described by Heikkilä et al. In fact, company A, which primarily seeking growth starts its digital business model innovation path by focusing on understanding and better serving their customer base. At the end of their path digitization of key activities, to deliver more flexible production processes leads to increased efficacy. In the second case, company B, which primarily driven by improved efficacy starts its business innovation path with digitization of their production activities and moving toward better value delivered to customers. In summary the approach of Heikkilä et al. provide a healthy rational for company to initiate their digitization. Through positive effects of first steps in their innovation paths, companies take more active role in adopting digital technologies to other aspects of their business models. Despite the divergent initial starting point both companies end up with serving both growth and efficiency strategy. This shows the importance of internal robustness of the business model during the process of its innovation. For instance, for company A, continuing to better serve their increasing customer bases called for a need to improve the efficiency of their key activity. Both case study companies regard the process of digitizing their business model as necessary to stay competitive. A completely digitized business model is not achievable at once for the SMEs. They have to prioritize the innovation projects to start with and pick the compatible technologies for their goals. In both matters, DIHs like the IMI Brandenburg had proved to be a great support to SMEs in those manners. Additionally, funds, e.g. from the European Commission can be selected and applied for in cooperation with the IMI Brandenburg to further enable digital business model innovations of SME.

6 Conclusion

This exploratory case study article aims to elaborate on iterative innovation steps that lead to BMI paths. In addition, through the two case studies the authors describe problems that SMEs faced and possibilities of assistance they found in a digital innovation hub. The method of analyzing case studies suits the exploratory nature of this research. The analysis of the two cases is not generalizable. However, the results strengthen the argument that business model innovation in SMEs is not one single step process but need several iterations. Those iterative steps need to be based on prior prioritization of primary goals of the company (growth or efficiency). Since funding and other resources for innovation are limited in SMEs, the companies need partners for successful innovation. Digital innovation hubs provide the key to a network of partners and expert knowledge. Since the two companies successfully pursue continuous innovation, they could be categorized as best practice examples. Thus, more in depth research should analyze the BMI process based on digital technologies in different companies. This includes a broader empirical basis as well as scales of digitization and BMI as a result.

This paper contributes to the scholarly debate on the nature of BMI by providing an empirical evidence of the dynamic approach adopted by SME to digitize their activities. For managers of SMEs, this paper provides a blue print of two successful paths to digital-based BMI.

References

1. Kunath, M., Winkler, H.: Integrating the Digital Twin of the manufacturing system into a decision support system for improving the order management process. Procedia CIRP **72**, 225–231 (2018). https://doi.org/10.1016/j.procir.2018.03.192
2. Aagaard, A., Presser, M., Beliatis, M., Mansour, H., Nagy, S.: A tool for internet of things digital business model innovation. In: 2018 IEEE Globecom Workshops (gc WKSHPS) (2018)
3. Foss, N.J., Saebi, T.: Fifteen years of research on business model innovation: how far have we come, and where should we go? J. Manag. **43**(1), 200–227 (2017). https://doi.org/10.1177/0149206316675927
4. Osterwalder, A., Pigneur, Y., Clark, T.: Business model generation: a handbook for visionaries, game changers, and challengers. [Amsterdam]: Alexander Osterwalder & Yves Pigneur (2010)
5. Müller, J.M., Buliga, O., Voigt, K.-I.: Fortune favors the prepared: How SMEs approach business model innovations in Industry 4.0. Technol. Forecast. Soc. Change **132**, 2–17 (2018). https://doi.org/10.1016/j.techfore.2017.12.019

6. Tangour, C., Gebauer, M., Fischer, L., Winkler, H.: Digital business model patterns of big pharmaceutical companies - a cluster analysis. In: Jallouli, R., Bach Tobji, M.A., Bélisle, D., Mellouli, S., Abdallah, F., Osman, I. (eds.) Digital Economy. Emerging Technologies and Business Innovation, Bd. 358, pp. 397–412. Springer International Publishing, Cham (2019)

7. Ammar, O., Chereau, P.: Business model innovation from the strategic posture perspective an exploration in manufacturing SMEs. Eur. Bus. Rev. **30**(1), 38–65 (2018). https://doi.org/10.1108/EBR-09-2016-0119

8. Marolt, M., Lenart, G., Maletic, D., Borstnar, M.K., Pucihar, A.: Business model innovation: insights from a multiple case study of slovenian SMEs. Organizacija **49**(3), 161–171 (2016). https://doi.org/10.1515/orga-2016-0015

9. Bouwman, H., Nikou, S., de Reuver, M.: Digitalization, business models, and SMEs: how do business model innovation practices improve performance of digitalizing SMEs? Telecommun. Policy **43**(9), 101828 (2019) https://doi.org/10.1016/j.telpol.2019.101828

10. Statistisches Bundesamt, „Anzahl der Industrieunternehmen in Deutschland nach Beschäftigtengrößenklassen und Bundesländern im Jahr 2017 (2019). Zugegriffen: Apr. 30, 2020. [Online]. Verfügbar unter. https://de.statista.com/statistik/daten/studie/311364/umfrage/industrieunternehmen-in-deutschland-nach-beschaeftigtengroessenklassen-und-bundeslaendern/

11. Kampe, C., Walter, A., Porep, D.: Arbeit 4.0 in Brandenburg. Zusammengefasste Ergebnisse zu Digitalisierungsniveaus - Beschäftigungseffekten - Arbeitsformen - Qualifizierungsbedarfen, Wirtschaftsförderung Land Brandenburg GmbH (2018). [Online]. Verfügbar unter: https://www.wfbb.de/de/system/files/media-downloads/wfbb_studie_digitalisierung.pdf

12. Massa, L., Tucci, C.L., Afuah, A.: A critical assessment of business model research. Acad. Manag. Ann. **11**(1), 73–104 (2017). https://doi.org/10.5465/annals.2014.0072

13. Taeuscher, K., Abdelkafi, N.: Visual tools for business model innovation: recommendations from a cognitive perspective. Creat. Innov. Manag. **26**(2), 160–174 (2017). https://doi.org/10.1111/caim.12208

14. Joyce, A., Paquin, R.L.: The triple layered business model canvas: a tool to design more sustainable business models. J. Clean. Prod. **135**, 1474–1486 (2016). https://doi.org/10.1016/j.jclepro.2016.06.067

15. Bucherer, E., Eisert, U., Gassmann, O.: Towards systematic business model innovation: lessons from product innovation management: towards systematic business model innovation. Creat. Innov. Manag. **21**(2), 183–198 (2012). https://doi.org/10.1111/j.1467-8691.2012.00637.x

16. Wirtz, B., Daiser, P.: Business model innovation processes: a systematic literature review. J. Bus. Models **6**(1), 40–58 (2018). https://doi.org/10.5278/ojs.jbm.v6i1.2397

17. Demil, B., Lecocq, X.: Business model evolution. in search of dynamic consistency. Long Range Plann. **43**(2–3), 227–246 (2010). https://doi.org/10.1016/j.lrp.2010.02.004

18. Andulkar, M., Le, D.T., Berger, U.: A multi-case study on Industry 4.0 for SME's in Brandenburg, Germany. In: Proceedings of the 51st Hawaii International Conference on System Sciences, pp. 4544–4553 (2018)

19. Cosenz, F., Bivona, E.: Fostering growth patterns of SMEs through business model innovation. A tailored dynamic business modelling approach. J. Bus. Res. S0148296320301594 (2020). https://doi.org/10.1016/j.jbusres.2020.03.003

20. Bianchi, C., Winch, G., Cosenz, F.: Experimenting lean dynamic performance management systems design in SMEs. Int. J. Product. Perform. Manag. **67**(7), 1234–1251 (2018). https://doi.org/10.1108/IJPPM-10-2017-0266

21. Lindgren, P.: Business model innovation leadership: how do sme's strategically lead business model innovation? Int. J. Bus. Manag. **7**(14) (2012). https://doi.org/10.5539/ijbm.v7n14p53

22. Gruber, H.: Proposals for a digital industrial policy for Europe. Telecommun. Policy **43**(2), 116–127 (2019). https://doi.org/10.1016/j.telpol.2018.06.003

23. Heikkila, M., Bouwman, H., Heikkila, J.: From strategic goals to business model innovation paths: an exploratory study. J. Small Bus. Enterp. Dev. **25**(1), 107–128 (2018). https://doi.org/10.1108/JSBED-03-2017-0097

24. Heikkilä, M., Bouwman, H., Heikkilä, J.: From strategic goals to business model innovation paths: an exploratory study. J. Small Bus. Enterp. Dev. **25**(1), 107–128 (2018). https://doi.org/10.1108/JSBED-03-2017-0097

25. European Commission: SBA Fact Sheet Germany". Zugegriffen: März 05, 2020. (2019) [Online]. Verfügbar unter. https://ec.europa.eu/docsroom/documents/38662/attachments/12/translations/en/renditions/native

26. Cavalcante, S., Kesting, P., Ulhøi, J.: Business model dynamics and innovation: (re)establishing the missing linkages. Manag. Decis. **49**(8), 1327–1342 (2011). https://doi.org/10.1108/00251741111163142

27. Arendt, L.: Barriers to ICT adoption in SMEs: how to bridge the digital divide? J. Syst. Inf. Technol. **10**(2), 93–108 (2008). https://doi.org/10.1108/13287260810897738

28. Rauch, E., Vickery, A.R., Brown, C.A., Matt, D.T.: SME requirements and guidelines for the design of smart and highly adaptable manufacturing systems. In: Matt, D.T., Modrák, V., Zsifkovits, H. (eds.) Industry 4.0 for SMEs, pp. 39–72. Springer International Publishing, Cham (2020)

29. Kampe, C., Walter, A.: Wirtschaft 4.0 in Brandenburg! Eine explorative Vorstudie der Wirtschaftsförderung Land Brandenburg GmbH (2017)

30. Solaimani, S., Heikkilä, M., Bouwman, H.: Business model implementation within networked enterprises: a case study on a finnish pharmaceutical project: business model analysis. Eur. Manag. Rev. **15**(1), 79–96 (2018). https://doi.org/10.1111/emre.12124

31. Gupta, R., Mejia, C., Kajikawa, Y.: Business, innovation and digital ecosystems landscape survey and knowledge cross sharing. Technol. Forecast. Soc. Change **147**, 100–109 (2019). https://doi.org/10.1016/j.techfore.2019.07.004

32. European Commission: What's in it for business - shaping Europe's digital future (2020)

33. Kilimis, P., Zou, W., Lehmann, M., Berger, U.: A survey on digitalization for smes in Brandenburg, Germany. IFAC-Pap. **52**(13), 2140–2145 (2019). https://doi.org/10.1016/j.ifacol.2019.11.522

34. Jans, R., Dittrich, K.: A review of case studies in business research. In: Case Study Methodology in Business Research, pp. 19–29. Elsevier, Boston (2007)

35. Dul, J., Hak, T.: Case Study Methodology in Business Research. Elsevier, Boston (2007)

36. Yin, R.K.: Case Study Research: Design and Methods, 4th edn. Sage Publications, Los Angeles (2009)

Pricing Digital Arts and Culture Through PWYW Strategies

A Reconsideration of the Ricardian Theory of Value

Racquel Antoun-Nakhle, Nizar Hariri, and Rim Haidar

Faculté de Sciences économiques, Observatoire Universitaire de la Réalité Socio-Economique, Université Saint-Joseph, Beirut, Lebanon
{racquel.nakhle,nizar.hariri}@usj.edu.lb, rimhaidar@hotmail.com

Abstract. Starting from empirical observations, this paper discusses the valuation problem of digital arts and culture, exploring the potentialities of value-based pricing strategies for digital arts markets by specifically examining the case of informal art market in Lebanon. In particular, the "Pay What You Want" (PWYW, also known as "Pay As You Wish") participative pricing strategy seems suitable for products that are largely dependent on untraditional motivations that contradict the rational choice theory, especially when the marginal costs are insignificant for producers or can't be calculated objectively for artistic and cultural products. We suggest that the PWYW pricing strategy in the digital art markets leads to a shift from price discrimination to value discrimination, when the value in use of reproducible digital arts is determining the exchange value, and the producers are capturing the maximum customers' surplus they are entitled to receive.

Keywords: PWYW · Use value · Pricing · Consumer fairness · Value discrimination

1 Introduction

In tandem with the expansion of digital markets of arts (digital books, photos, music, movies, etc.), suppliers and creators are initiating new ways to differentiate and personalize their products. Producers are increasingly adopting value-based pricing strategies where customers are actively participating in the pricing process in order to ad-just the supply to various segments of consumers with specific tastes and preferences. In particular, the "Pay What You Want" (PWYW) participative pricing strategy al-lows customers to choose the price they would like to pay when buying a product or service and the supplier should passively accept any allocation.

Nevertheless, according to the neoclassical model, a rational customer will tend to pay the lowest price allowed by the vendor, which in the case of the PWYW could go as low as zero. Yet, most empirical studies show that not all consumers will adopt this minimum pricing strategy, and the PWYW participative pricing strategy expands retailers' revenues and profits (Kim et al. 2010; Kunter 2015; Viglia et al. 2019). In-deed, if well implemented, value-based pricing seems to allow the producer to de-crease the opportunity costs and to capture a greater share of consumers' surplus.

© Springer Nature Switzerland AG 2020
M. A. Bach Tobji et al. (Eds.): ICDEc 2020, LNBIP 395, pp. 166–176, 2020.
https://doi.org/10.1007/978-3-030-64642-4_14

That said, what would encourage rational suppliers or creators aiming to maximize their profits to adopt the value-based pricing strategies? How can a PWYW strategy effectively increase the producers' surplus, while allowing the consumer to pay the lowest amount? If vendors are concretely increasing their profits, could this pricing strategy be a disguised form of price discrimination, leaving the customers with the false impression of saving money? Moreover, in the PWYW strategy, what could encourage consumers to pay a positive price for a product they could acquire for free?

Pricing is one of the major determinants of profitability and economic growth, yet it remains a theoretical enigma for economists and analysts. Valuation and pricing in the digital era are major challenges for economists especially for artworks and paintings; In fact, the dematerialization, the technical and digital reproducibility, as well as the digital sales of art are reviving with them the old debate between the proponents of neoclassical utility-value and valuation by rarity on one hand, and the Ricardian and Neo-Ricardian theories of labor value on the other hand. In this sense, this paper offers some reconsideration of the theory of value and the value paradox of artworks (tangible or intangible) in the context of their digital reproducibility. It is well known that Ricardo (1821) distinguished between two types of goods: reproducible and non-reproducible goods. The former have an intrinsic value that is independent from its use value, since it could be objectively calculated according to the value of the incorporated labor, when the cost of production determines the exchange value. The latter are rare products that could not be reproduced by labor, such as original paintings or sculptures, coins or stamps, which exchange value will be determined by their use value.

Yet, digital art seems to contradict both neoclassical and Ricardian theories of value. Indeed, it is infinitely reproducible by definition, with average and marginal costs quickly close to zero. The demand functions also contradict the traditional logic of neoclassical models since the relative scarcity of the artwork has almost no influence on the willingness to pay, unlike paintings, canvases, or sculptures whose uniqueness and authenticity are major determinants of value.

Thus, traditional valuation models, namely cost production pricing as well as market competitiveness pricing, seem obsolete in the era of digital distribution of digital arts. Indeed, production costs pricing is strictly impossible for digital artworks, since the creative investment in the art piece can't be estimated by time, efforts, or materials used during the creative process. Similarly, following the logic of supply and demand, a pricing strategy by market competition could be sub-optimal in large segments of the market, or simply forbidden for a big number of products for which a market does not even exist. In some extreme cases, like illegal downloading of music on the net, each new consumer considers that it is legitimate to benefit freely from a digital product since the producer is not incurring any additional cost. However, on the macroscopic level, the producer surplus can be entirely extorted by the consumers, and in some extreme cases a legal framework for this type of markets wouldn't even exist, since the producers will not have any incentives to offer their products.

Despite unlimited technological reproducibility, digital art retains an emotional value for art lovers. For some fans, buying a reproduction of their favorite artist is a glamorous experience for which they are willing to pay much more than its cost. The para-dox of the value of digital art is linked to the sub-optimality of traditional pricing methods. Indeed, any price that would be fixed above the marginal cost (which is close to zero) will exclude a wide range of consumers who won't have the willingness to pay this over-estimated price, leading to an underutilization problem. Nonetheless, if a consumer decides to pay the product at its marginal cost, which is always close to zero, the market will suffer from a production problem, since artists and creators won't be able to secure any decent payment for their creative work.

In the case of value-based pricing of digital arts, price fixation generally involves a negotiation between creators and potential buyers; the bargaining concerns both the artistic value of the product and its exchange value. In this sense, pricing strategies based on the value in use for customers seem to provide solutions to the paradox of value of the digital art, since the value of an artwork is actively determined by the buyers themselves. This pricing strategy seems perfectly suited to certain market structures, especially for cultural or creative products, artworks and live performances. Indeed, the Ricardian evaluation of the price by the production costs is in all cases impossible, since one cannot estimate the value of a painting or a concert by the quantity of painting or energy involved in the process of production. Similarly, market competition pricing is only possible for artworks for which a formal market exists, but it could be ineffective for young creators or informal artistic productions, where a formal or a legal market does not even exist, as in the case of street arts and free public performance venues. For example, after a free theatrical performance or a street concert, it is common to see the artist circulating between spectators with a hat in their hand, in order to collect remuneration based on the logic of PWYW.

In particular, the PWYW pricing strategy seems suitable for digital arts that are largely dependent on the emotions they procure for consumers, especially when the marginal costs are insignificant for producers. Starting from empirical observations and some case studies from the Lebanese art market, the next section explores the valuation problem of digital cultural products. Based on an extensive literature review, the third section aims to understand the motivations of creators and art consumers engaged in a participative pricing strategy. Section four presents a practical framework, exploring the potentialities of value-based pricing strategies for digital art markets. The discussion develops some theoretical reconsiderations of the Ricardian theory of value. It suggests that the PWYW pricing strategy in the digital art markets leads to a shift from price discrimination to value discrimination, when the value in use of re-producible digital arts is determining the exchange value, and the producers are capturing the maximum customers' surplus they are entitled to receive.

2 Empirical Examples: Old and New Methods of PWYW

In October 2007, the famous British band Radiohead sold its new album In Rainbows via a website designed for this purpose by asking customers to "pay what they want" to acquire it (Van Buskirk 2007). Since the average price paid by the consumers (USD 6) was inferior to the standard album price, Radiohead's strategy was criticized, and even listed in the "101 Dumbest Moments in Business" by the Fortune Magazine in 2007. Nevertheless, when paying an average price of 6 USD, a great number of fans freely chose to pay a positive endowment, while they could have downloaded (legally or illegally) the album without paying any fees.

Radiohead's marketing strategy in 2007 wasn't new. However, we believe that the band's implementation of the PWYW strategy led to its ubiquity.

Among the pioneer studies about the PWYW participative pricing strategy was that of Steiner (1997). The latter led an empirical study concerning the effects of an indirect PWYW pricing strategy in a museum. He found that letting visitors enter a museum without entrance fees encourages them to buy more souvenirs and food from the museum's cafeteria and souvenirs shop. Yet, he found that this increase in the sales of the souvenirs shop and the cafeteria is not optimal since fixed entrance fees are more profitable to the museum. The reason underlying the latter is intuitive and predicated on visitors' rationality. According to Rao and Sieben (1992), an entrance fee per unit just serves as a reference to the visitors who rationally tend to discount this reference.

Der Wiener Deewan, a famous restaurant that serves Indian and Pakistani dishes in Vienna ranked 4.7/5 on Google.com is found on the PWYW strategy. The restaurant's all-you-can-eat and pay-what-you-want buffet as well as play-what-you-want jam attracts visitors from all over the world. Its success owed to its original participative pricing strategy ensured the restaurant's continuity and attractiveness over the years.

In 2014, Motto, a restaurant in Lebanon (Mar Mikhael) opened its doors to the clients and let them pay what they want for the offered food. In order to ensure the persistence of the business, a referential minimum price of 9,000 LBP was suggested by the owners. Despite this, as shown on Zomato.com, the restaurant doesn't use the PWYW strategy anymore.

Some travel agencies also use or used the PWYW pricing strategy. Atrápalo, a Spanish travel agency, launched its PWYW deal ("El trato de Atrápalo") in 2009 which consisted in offering bundles to the consumers under PWYW conditions. Their strategy proved to be efficient and increased the agency's profits (León et al. 2012). An-other travel agency based in the United States, bonvoyadventuretravel.com, also lets its clients price its deals.

Greentoe.com, a website that sells equipment for photography and music also applies the PWYW pricing strategy. However, it displays a referential minimum price for the consumers in order to make their pricing decisions converge to a fair price.

Creative Space Beirut is a non-lucrative fashion design school in Lebanon that uses the PWYW participative pricing strategy when selling the creations of its students. The Creative Space Beirut is a unique case of a free yet sustainable fashion school for underprivileged students, as stated in the Creative Economy Report made by the UNESCO in 2013. During fashion shows taking place in the Gulf and in Lebanon, spectators have the choice to privately write their prices in an envelope in order to purchase a unique Haute-Couture artwork. It seems that this strategy was successful, since these exhibitions were the main source of funding for this non-lucrative school.

M. is a young Syrian painter who was displaced in Lebanon at the beginning of the war in Syria in 2011. M. was quickly displayed in the main galleries of the Lebanese capital, benefiting from the rapid growth of the art market in Lebanon, and the rise in the global demand for arts in a context of refugees. Coming out of anonymity, M. has seen his popularity rating and his quotation in private sales multiplied by 10. In the main art galleries, he now sells his paintings on the Lebanese market at an average price of 9,000 USD, while some of his paintings travel in exhibitions around the globe for international exhibitions from which the artist himself might be sometimes excluded because of his refugee status.

However, M. sells paintings, lithographs and reproductions by other channels. Like many young artists, M. exhibits his products on the Internet, on specialized sales sites, and on social media. Fans can contact him directly about certain paintings, negotiate the price online, or visit him in his studio to conclude a deal.

Except for paintings distributed via formal infrastructures and the price of which is contractually fixed by curators or galleries (with a range varying between 5000 and 17000 USD), these artworks have no predetermined price. Paintings sold directly by the artist have a price negotiated by buyers who assess the amount they are willing to pay, giving the artist the right to accept or refuse.

3   Why Pay Less if You Can Pay More?

For digital cultural products, each consumer will logically purchase one unit of the goods over a given period. The firm's demand curve is an arrangement of the consumers according to their reservation prices - i.e. the maximum price each consumer is willing to pay. For these goods, knowledge of the demand curve means that the firm knows that the top part of the demand curve, namely the "tall head", is made up of those consumers willing to pay a lot for the one unit they will purchase, while the bottom part of the demand curve, namely the "long train", is made up of those willing to pay a little.

In his "long tail of products", Anderson (2006) theorized the latter. He argues that using the PWYW pricing strategy in digital markets offers e-businesses an unprecedented opportunity to increase their revenues. E-businesses selling non-reproducible products have an advantage over traditional stores. For instance, the latter face storage limitations and thus only tend to sell "Bestsellers". E-retailers, on the contrary, can gain additional revenues by selling less-popular products since they don't have inventory constraints.

Reisman et al. (2019) adapted Anderson's (2006) idea to the PWYW pricing strategy. They illustrate the latter in their "long tail of customers" which suggests that e-businesses implementing the PWYW pricing strategy shouldn't base their profits anticipations solely on the revenues derived from consumers who are willing to pay less than the standard price (long tail), but also on the revenues derived from consumers who are willing to pay more than the standard price (tall head).

E-retailers benefit from an additional consumer surplus from tall head consumers who pay more than the standard price, but they also benefit from long tail consumers who perceive a "use value" in the product and pay more than the marginal cost. In this way, e-suppliers "secure lost revenues" (Reisman et al. 2019). Indeed, by paying a sum lower than the estimated price, customers engaged in pricing by the value in use do not reduce the producers' surplus. They just allow them to capture the maximum consumer surplus they are entitled to receive.

Suppliers usually attract consumers who have different price elasticities through different offers such as special deals, discounts, coupons, bundles, etc. However, these methods are typically applicable on the short run, for a specific time period. Moreover, retailers often initiate such offers without any planning or strategy. In digital markets, such offers, even unorganized, can be extended for longer periods since the marginal costs of selling additional products is insignificant.

Yet, there may be tall head consumers who are willing to pay more for a product, but don't practically do so because they're not price sensitive. In such case, e-retailers capitalize on the consumers' surplus by adding features to the product in order to make it more "valuable" for the consumers. For instance, products containing advertisements may be more valuable for the consumers (Hungenberg et al. 2008).

The dictator game is a perfect illustration of the PWYW strategy. If the dictator (the buyer) is willing to offer a very low price, it may be more relaxing and less morally stressing to acquire the product via the internet. In fact, they won't be face-to-face with the supplier, which could guaranty a sense of anonymity and allow a better decision making. Therefore, when choosing what one would like to pay to acquire a given product, the consumer sitting behind the screen won't be submitted to any social stigma or peer-pressure, and the price chosen will tend to reflect their own preferences. As shown by Elster (2009), disinterested yet not irrational choices could be part of our daily life (such as in the act of voting), and some forms of selflessness are shaped in the double anonymity of an act of giving. Thus, each time a Radiohead's fan chooses to give any positive endowment (>1$), this decision could be interpreted as being motivated by a double disinterested choice: the positive endowment is not known from the "receiver" (the band doesn't personally know the "proposer"); it is not known from the public (since the buyer can't take advantage of this generous act).

Thus, economic behavior can't be solely based on economic rationality, especially in the art markets, when agents are also driven by emotions, talent, creativity and aesthetic judgments.

For this reason, it is crucial to focus primarily on the motivations underlying the PWYW pricing strategy. For instance, some consumers may value their relationship with the supplier, which makes them pay a fair price in order not to damage the sup-plier's surplus (Marett et al. 2012). Furthermore, some consumers may perceive the supplier's offer to pay what they want as an act of generosity and will act out of reciprocation somewhat to "reward" the supplier for their kind behavior (Kim et al. 2009). For example, if the supplier offers samples of their product to the consumer before asking them to pay what the latter wants, the consumer will perceive the supplier's act as generous and will, therefore, instinctively want to reciprocate it by paying a positive price to buy the product. Hence, the relationship between buyers and sellers is governed less by market exchange norms than by social exchange ones.

Whilst some consumers may be oriented to act out of sympathy, others will pay fairly in order to fulfill their own interests, which is rational. In fact, such consumers may pay fairly in order to ensure the perpetuation of the supplier's offer (Gneezy et al. 2012).

Yet, e-suppliers don't rely passively on the consumers' fairness assumption. Indeed, they may elaborate strategies to implicitly motivate the consumer to pay a fair price. In fact, prior to deciding the price to pay, consumers ask their fellows about the price they would pay to acquire such a product. They may also make researches on the web in order to determine the price. For this reason, e-retailers display a reference price for the product to adjust the consumers' anticipations of the price (Shampanier et al. 2007). For instance, Greentoe.com, a famous website implementing the PWYW pricing strategy displays the lowest price online of the product in order to help consumers choose a fair price.

E-suppliers may, also, display implicit or explicit cues about the costs they incur in order to emit a "signal" that would attract fair consumers (Schons et al. 2013).

Empirical literature, also, focused on the importance of timing when implementing a PWYW pricing strategy. Gautier and van der Klaauw (2012) stipulate that asking the consumers to pay what they want after their engagement in the transaction will lead them to pay a higher price for a product than asking them to do so before engaging in the transaction.

4 Practical Framework

We suggest a practical framework for artistic products sold online under the PWYW participative pricing strategy (see Fig. 1). Our framework is inspired by the FairPay framework (Reisman et al. 2019).

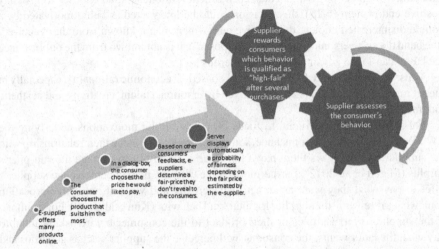

Fig. 1. A new dynamic framework for the implementation of PWYW for digital arts and culture.

We consider that painter M. establishes a website where he would sell his paintings. M. who becomes an e-supplier, will display pictures of his paintings on the website. Knowing that he acquired popularity on social media and regular markets by negotiating his paintings' prices with buyers, consumers will be interested in visiting his website and buying or rebuying paintings.

When a consumer visits the website, they will choose the painting for which they attribute, logically, the highest use value. In other words, they will choose a painting for which they would logically pay, at least, a fair price.

Like in the FairPay framework (Reisman et al. 2019), the e-supplier (M.) has several indirect interactions with each consumer in order to assess their behavior and rate it as "unfair", "fair" or "high-fair". Based on this assessment, M. rewards consumers which behavior is qualified as "high-fair" after several purchases by, for example, sending them a signed painting. In that way, consumers are motivated to pay a highly fair price.

Yet, on the short run, when making their first purchase, or even their third one, consumers won't be motivated to act fairly or highly fair since they were never rewarded by the e-supplier before. For this reason, the supplier has to "nudge" consumers to pay a fair price. In behavioral economics, one should never ignore the power of nudges that refer to "any aspect of the choice architecture that alters people's behavior in a predictable way without forbidding any options or significantly changing their economic incentives" (Thaler and Sunstein 2008).

For instance, our painter M. could indeed display a dialog box in which the consumer chooses a price (based on his estimation of the product's use value). However, in order to incentivize the consumer to pay at least a fair price on the short run, M. has to display, for each price the consumer types in the dialog box, a "probability of fair-ness" that would nudge the consumer and help them converge towards a reference price. The latter is estimated by M. according to feedbacks of consumers as well as viewers on the paintings' appeal, originality and quality. The "probability of fairness" displayed to the consumer is calculated by the distance between the price chosen by the consumer and the reference price calculated by the e-supplier. If the probability of fairness is 1, then the price set by the consumer corresponds exactly to the reference price. However, except if they see a probability of 1 for the price they set, the consumer won't know the reference price set by the e-retailer.

This display is supposed to be effective in sensitizing the consumer. In fact, Christopher and Machado (2019) stated that design variations in PWYW are factors that influence consumers' payments.

Displaying such probability to the consumer does not only nudge them, but it also awakens their subconscious "social preferences". Alexandrov et al. (2013) insisted on consumers' positive social image: they can adopt prosocial behavior to gain prestige or favorable reputation especially in visible payments.

We believe that using such framework for artistic goods is efficient. However, empirical tests must be made to prove its effectiveness.

5 Discussion and Conclusion

Pricing based on the value in use for customers is mainly linked to the value that a product or a service provides to various segments of customers, letting them actively set the price (thus, the intrinsic value of the product). This type of pricing, if properly adopted, can have positive correlations with profits and it may ensure a sustainable sales growth.

In contrast to pricing strategies based on production costs (the Ricardian theory of labor value), pricing strategies based on use value estimated by consumers appear today as new ways to capture a higher share of consumers surpluses, thus providing a solution that is both legal and competitive to the problem of price discrimination.

Indeed, sometimes it seems completely illegitimate (or illegal) for a firm to charge differentiated prices according to the willingness to pay of various consumers. On one hand, the moral economy of price discrimination is only justified in terms of price discounts, i.e. a sacrifice of the producer profits in order to lower its prices to potential buyers who would otherwise be excluded. For example, student prices (for movies, plays or museums) would be morally justified by the democratization of arts and culture, where premiums towards the seniors or the experts would be guilty of cream-skimming markets, and perceived as illegitimate forms of discrimination. On the other hand, the practice of a single price for different segments of consumers who do not have the same preferences and purchasing powers is part of logic of substitution from one market segment to another. By charging a higher price for a coffee perceived as being of better quality, Starbucks sacrifices part of its customers to MacDonald's, and tries to capture a loyal clientele with a higher purchasing power. What if the producer could adapt its products to every segment of its clientele? If this type of pricing is not easy to implement in the coffee industry, it seems perfectly adjusted for informal art sales.

As a matter of fact, artists implementing a fixed price strategy expose themselves to the risk of losing a significant market share of consumers with a lower willingness to pay.

Therefore, the value-based pricing strategy allows these two problems to be solved simultaneously, by establishing individualized pricing determined by the consumer's assessment while providing a moral justification for the overpricing.

In fact, pricing according to the value in use for the customer can only be prejudicial to the seller if the product has all the characteristics of a rival merchandise, a labor intensive product that is only reproducible by increasing labor and capital, and acquiring an intrinsic value strongly linked to its production costs, due to the existence of a decreasing marginal productivity (notably for the labor factor). However, in the case of digital art, the traded merchandise has none of these characteristics: It is non-rival, and infinitely reproducible without significant costs, with increasing returns to scales. In this case, the value of the object is completely independent of its cost of production, for both parties. For the artist, working time does not respond to the energy model of increasing marginal disutility of labor, since the creative process is not experienced according to the rules of constrained optimization (minimize effort or maximize output). Similarly, for art lovers, the value of the product is rarely linked to the effort invested by the artist, and it depends more significantly on the creative investment and the emotion produced by the artwork. In both cases, we are not facing a zero-sum game, where what is won by one is lost by the other. The assumption that a vendor will suffer from a loss every time the consumer is paying a price below the "fair price" is simply incorrect.

Particularly for personalized, signed or autographed artworks by their creators, pricing strategies depend on an estimation of the value by the potential consumers, leaving creators or e-suppliers with three possibilities. Either the value estimated by the future buyer is equivalent to the supplier's estimation, or it is higher or lower. Only in the latter case, the seller seems to incur a loss. The e-supplier knows very well the time and talent that they invest in each of their creations and the reproductions signed have fluctuating values depending on each client. It would be rational for any art creator to establish between them and their fans a pricing strategy based on subjective valuation, when pricing is negotiated according to the fluctuating value in use for the client. For instance, the production of a signed poster does not incur additional costs for a painter, since the marginal cost tends to zero. In the meanwhile, the acquisition of a signed copy by the customers ensures a symbolic gratification that they are willing to pay according to their subjective scales.

The main contribution of digital technologies to pricing mechanisms is the transition from fixed pricing processes to "value-based pricing" processes. The latter underline a long-term oriented strategy residing in personalized prices that become more and more accurate and fair the longer the supplier's relationship with the consumer is.

The PWYW participative pricing strategy can be said a disguised, first degree discrimination or personalized pricing scheme. Yet, it, also, leaves the supplier with a better understanding of the consumer's sense of value. This dynamic participating pricing opens space for a bargaining between customers and suppliers regarding the intrinsic value of the product and its value in use. The customer offers a price that reflects value in-use (Reisman et al. 2019), and if accepted by the producers, this participative pricing would enhance value to both parties. In such case, the transaction will shift from price discrimination into "value discrimination".

References

Alexandrov, A., Lilly, B., Babakus, E.: The effects of social-and self-motives on the intentions to share positive and negative word of mouth. J. Acad. Mark. Sci. **41**(5), 531–546 (2013)

Anderson, C: The Long Tail: Why the Future of Business is Selling Less of More. Hachette Books (2006)

Christopher, R.M., Machado, F.S.: Consumer response to design variations in pay-what-you-want pricing. J. Acad. Mark. Sci. **47**(5), 879–898 (2019). https://doi.org/10.1007/s11747-019-006 59-5

Elster, J.: Le désintéressement. Seuil, Paris (2009)

Gautier, P.A., Klaauw, B.V.D.: Selection in a field experiment with voluntary participation. J. Appl. Econ. **27**(1), 63–84 (2012)

Gneezy, A., Gneezy, U., Riener, G., Nelson, L.D.: Pay-what-you-want, identity, and self-signaling in markets. Proc. Natl. Acad. Sci. **109**(19), 7236–7240 (2012)

Hungenberg, H., Enders, A., Denker, H.P., Mauch, S.: The long tail of social networking: revenue models of social networking sites. Eur. Manag. J. **26**(3), 199–211 (2008)

Kim, J.Y., Natter, M., Spann, M.: Pay what you want: a new participative pricing mechanism. J. Mark. **73**(1), 44–58 (2009)

Kim, J.Y., Natter, M., Spann, M.: Kish: Where customers pay as they wish. Rev. Mark. Sci. **8**(2) (2010)

Kunter, M.: Exploring the pay-what-you-want payment motivation. J. Bus. Res. **68**(11), 2347–2357 (2015)

León, F.J., Noguera, J.A., Tena-Sánchez, J.: How much would you like to pay? Trust, reciprocity and prosocial motivations in El trato. Soc. Sci. Inform. **51**(3), 389–417 (2012)

Marett, K., Pearson, R., Moore, R.S.: Pay what you want: an exploratory study of social exchange and buyer-determined prices of iProducts. Commun. Assoc. Inform. Syst. **30**(1), 10 (2012)

Rao, A.R., Sieben, W.A.: The effect of prior knowledge on price acceptability and the type of information examined. J. Consum. Res. **19**(2), 256–270 (1992)

Reisman, R., Payne, A., Frow, P.: Pricing in consumer digital markets: a dynamic framework. Aust. Mark. J. **27**(3), 139–148 (2019)

Ricardo, D.: On the Principles of Political Economy. J. Murray, London (1821)

Schons, Laura Marie., Rese, Mario., Wieseke, Jan., Rasmussen, Wiebke., Weber, Daniel, Strotmann, Wolf-Christian: There is nothing permanent except change—analyzing individual price dynamics in "pay-what-you-want" situations. Mark. Lett. **25**(1), 25–36 (2013). https://doi.org/10.1007/s11002-013-9237-2

Shampanier, K., Mazar, N., Ariely, D.: Zero as a special price: The true value of free products. Mark. Sci. **26**(6), 742–757 (2007)

Steiner, F.: Optimal pricing of museum admission. J. Cult. Econ. **21**(4), 307–333 (1997)

Thaler, R.H., Sunstein, C.R.: Nudge: improving decisions about health. Wealth Happiness **6** (2008)

UNESCO: Creative economy report 2013: Special edition: Widening local development pathways. UNCTAD (2013)

Van Buskirk, E.: ComScore: 2 Out of 5 Downloaders Paid for Radiohead's 'In Rainbows' (Average Price: $6)', WIRED, 5 November (2007)

Viglia, G., Maras, M., Schumann, J., Navarro-Martinez, D.: Paying before or paying after? Timing and uncertainty in pay-what-you-want pricing. J. Serv. Res. **22**(3), 272–284 (2019)

Author Index

Printed in the United States
by Bookmasters

Printed in the United States
By Bookmasters